Date Due

MAY 10 1973			
SEP 10 1973			

The Science and Technology of Superconductivity
Volume 2

The Science and Technology of Superconductivity

Proceedings of a summer course held August 13-26, 1971,
at Georgetown University, Washington, D. C.

Edited by
W. D. Gregory
and
W. N. Mathews Jr.
Department of Physics
Georgetown University
Washington, D.C.

and

E. A. Edelsack
Office of Naval Research
Washington, D.C.

Volume 2

PLENUM PRESS • NEW YORK–LONDON • 1973

Library of Congress Catalog Card Number 72-77226
ISBN 0-306-37632-6

© 1973 Plenum Press, New York
A Division of Plenum Publishing Corporation
227 West 17th Street, New York, N. Y. 10011

United Kingdom edition published by Plenum Press, London
A Division of Plenum Publishing Company, Ltd.
Davis House (4th Floor), 8 Scrubs Lane, Harlesden, London, NW10 6SE, England

All rights received

No part of this publication may be reproduced in any form without
written permission from the publisher

Printed in the United States of America

CONTENTS OF VOLUME 2

Contents of Volume 1 vii

PART III: TECHNOLOGICAL APPLICATIONS

Superconducting Power Transmission 433
    R. W. Meyerhoff

Application of Superconductivity in Thermonuclear
        Fusion Research 459
    A. P. Martinelli

Application of Superconductors to Motors and
        Generators 483
    Joseph L. Smith, Jr.

Superconducting Coils 497
    Z. J. J. Stekly

Physics of Superconducting Devices 539
    Bascom S. Deaver, Jr.

Superconductivity in DC Voltage Metrology 565
    T. F. Finnegan

Electric and Magnetic Shielding with Superconductors . . . 587
    Blas Cabrera and W. O. Hamilton

Superconductive Computer Devices 607
    J. Matisoo

Superconductors in Thermometry 625
    J. F. Schooley

Millimeter and Submillimeter Detectors and
        Devices 631
    Sidney Shapiro

Magnetometers and Interference Devices 653
    Watt W. Webb

PART IV: PANEL DISCUSSION

The Scientific, Technological, and Economic
      Implications of Advances in
      Superconductivity 681
    Edited by: W. N. Mathews Jr., W. D. Gregory,
        and E. A. Edelsack

PART V: CONCLUSIONS AND SUMMARY

The Technological Implications of Superconductivity
      in the Next Decade 719
    D. N. Langenberg

A Summary of the Course 735
    J. Bostock

PART VI: APPENDICES

1. Program 757

2. Invited Speakers 765

3. Participant List 767

4. Problems 771

Author Index 779

Subject Index 801

CONTENTS OF VOLUME 1

Contents of Volume 2   ix

PART I: INTRODUCTION AND BACKGROUND

Fundamentals of Superconductivity   5
    E. A. Edelsack

Phenomenological Theories of Superconductivity   25
    W. D. Gregory

Elements of the Theory of Superconductivity   71
    W. N. Mathews Jr.

Josephson Effect in a Superconducting Ring 149
    F. Bloch

Time-Dependent Superconductivity 163
    D. J. Scalapino

Refrigeration for Superconducting Devices 185
    Robert W. Stuart

PART II: SUPERCONDUCTING MATERIALS

Experimental Aspects of Superconductivity:
        Editors' Note 209

Superconductivity in Very Pure Metals 211
    W. D. Gregory

T_c's -- The High and Low of It 263
    Bernd T. Matthias

The Metallurgy of Superconductors 289
    Robert M. Rose

Superconducting Intermetallic Compounds -
        The A15 Story 333
    Robert A. Hein

Theory of Superconducting Semiconductors 373
    C. S. Koonce

Enhancement Effects: Theory 389
    C. S. Koonce

Enhancement Effects 405
    J. F. Schooley

Author Index xi
Subject Index xxxiii

VOLUME 2

Part III
Technological Applications

SUPERCONDUCTING POWER TRANSMISSION *

R. W. Meyerhoff

Union Carbide Corporation, Linde Division, Tarrytown
Technical Center, Tarrytown, New York 10591

At the present time it is predicted that the electric utilities will need to double their capacity every ten years. This rate of growth, about 8% per year, is not in itself awe-inspiring until one considers the present size of the electric industry. The estimated demand in the United States for the year 1990 is in excess of 10^9 kW. In order to meet this demand, the electric utilities will have to install approximately 7.5 x 10^8 kW of additional generating, transmission, and distribution capacity at an estimated cost of 300-350 billion dollars.

The availability of low cost and more efficient means of transporting and distributing large blocks of power may be the most significant factor which will determine the ability of the electric power industry to meet these rapidly growing demands. This is due to several factors. First, in place of the 3,000 plants in existence today, most of the new generating capacity installed in the next 20 years is expected to come from 250 huge plants, each of which will have an installed capacity of 2,000-3,000 mVA.[1] Kusco[2] has predicted that some of the plants built between 1980 and 1990 will have capacities up to 10,000 mVA. Kusko's conclusions are supported by the predictions of Boesenberg and Zanona[3], presented in Figure 1, for the future growth of a system having a 1970 peak load of 10,000 mVA. Boesenberg and Zanona predict that, with the increase in size of generating stations, there will be a concomitant increase in the capacity of transmission lines. Second, although at 345 kV, underground cables, including right-of-way, cost 15 to 16 times as much as overhead lines in suburban areas and 18 to 19 times as much in rural areas[4], both the increasing cost of right-of-way and the increasing public opposition to overhead power lines will necessitate an increase in the use of underground power

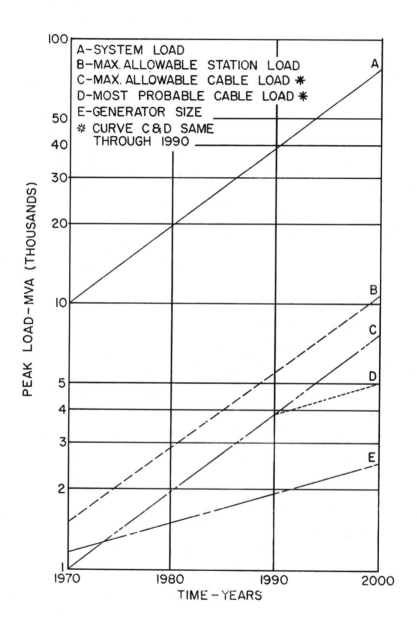

Figure 1. Peak load forecast for a 10,000 mVA system (Reference 3)

lines. For these reasons the electric utilities will be faced with the problem of transporting, via undergound cables, blocks of power even larger than those now carried by overhead lines. It is generally agreed that the present technology of underground power transmission is not capable of both efficiently and economically meeting anticipated requirements and a new technique for transporting power underground is needed.

A number of alternatives to present methods of underground power transmission have been proposed. Some of these alternatives which have been recently reviewed by others include microwave power transmission,[5,6] vacuum insulated cables operating at liquid nitrogen temperatures (77°K),[7] resistive cryogenic transmission lines,[8] and a combination in a single envelope of a cryogenic power transmission line and a liquefied natural gas transmission line.[9] Boesenberg and Zanona[3] have reviewed some of these new concepts for underground power transmission and have estimated both the power ratings at which of these systems would be economical as well as the year each of these systems would be available. Their conclusions are summarized in Figure 2. On the basis of these results, as well as other similar analyses, I believe that superconducting transmission lines hold the greatest promise for efficiently and economically meeting the requirements anticipated for the years following the mid-1980's. In the remainder of this paper, I have attempted to review the present state-of-the-art of superconducting underground power transmission.

HISTORY OF SUPERCONDUCTING POWER TRANSMISSION

The decade of the 1980's will, I believe, be characterized in part by the introduction of superconducting power lines, while the 1970's will be the decade in which the major portion of the engineering development and testing will be completed. Before beginning to assess the present state-of-the-art in superconducting power transmission, it is of value to review the work done during the early 1960's.

Superconductors may be divided into two classes, type I and type II. The type I superconductors, for example, lead, tin, and indium, have been known for many years. The use of a type I superconductor for power transmission was considered in some of the early studies described below. However, it is now generally agreed that the relatively low critical temperatures of the type I superconductors coupled with the fact that their low critical magnetic field, H_c, requires the use of large diameter conductors in order to maintain the surface magnetic field below H_c preclude their use for power transmission. Type II superconductors, for example, NbZr, NbTi, and Nb_3Sn, which were discovered in the 1950's, are

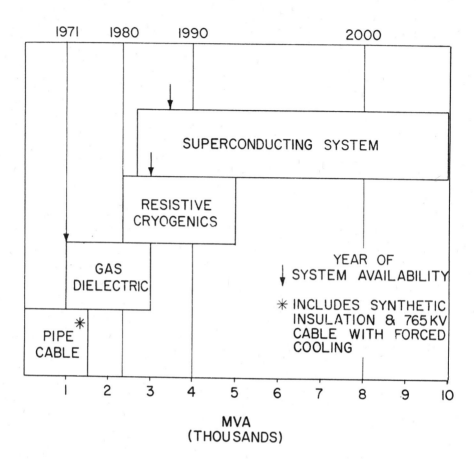

Figure 2. Capabilities of new underground transmission system (Reference 3)

characterized by relatively high critical temperatures and their ability to carry large currents in very high magnetic fields. Unfortunately, these highly irreversible type II superconductors have losses when placed in ac magnetic fields. For this reason, type II superconductors, with the exception of pure niobium, are considered to be applicable only for dc power transmission.

In 1961, McFee[10] was one of the first to seriously consider superconducting power transmission, although he did not consider the over-all system economics. McFee proposed the use of a 3 cm diameter Nb_3Sn conductor in a dc line operating at 200 kV with a rating in excess of 100,000 mVA as well as a 600 mVA single-phase ac cable using coaxial lead conductors having diameters of 3 and 6 cm.operating at 3,000 A. A coaxial lead cable was also discussed by Atherton[11] in 1963. Klaudy[12] in 1965 discussed the use of NbZr in an ac transmission line; however, like both Atherton and McFee, he did not provide cost estimates for the line.

One of the first studies to include an economic analysis was that of Gauster, Freeman, and Long[13] who proposed a 10 GW line operating dc at 150 kV, using hard niobium tubes as the conductors. The cost of such a line 1,000 miles (1 mile = 1.6 km) long, including the converters at each end of the line, was estimated to be $1,240 million ($124 per mVA-mile). This cost plus the operating cost was stated[13] "to compare well with similar calculations for 700 kV ac and 'conventional' ± 500 kV dc power lines." In 1967, Garwin and Matisoo[14] considered a 100,000 mVA dc line similar to that proposed earlier by McFee.[10] They concluded that a 625 mile line of this capacity operating at 200 kV and using Nb_3Sn as the superconductor would cost, including the converters, $806 million (∿ $13 per mVA-mile).

Because of the uncertainty about the magnitude of the ac losses in superconductors, the studies by Gauster, Freeman, and Long[13] and by Garwin and Matisoo[14] were restricted to economic analysis of dc lines. While the costs of these lines were attractive, the lines were of extremely high capacity and considerable length. Since over 25% of the cost of these dc lines was in the conversion equipment, the cost per unit length would increase very rapidly as the line lengths were reduced.

More recent studies of dc power transmission by both Voigt[15] and Delile[16] have shown superconducting dc lines to be competitive at much lower power levels than those considered by Gauster et al[13] and Garwin and Matisoo.[14] Voigt predicts a break-even point to be in the range 2-3 GW while Delile finds the superconducting line may be competitive at power levels as low as 1 GW. While these power levels are reasonable, these cost estimates did not include the conversion equipment. Thus, in the absence of a marked decrease in the cost of conversion equipment, for most power systems, the dc superconducting line cannot compete with an ac line except for very long lines. This assumes, as is discussed in the following section, that the ac losses in pure niobium do not prohibit the use of this type II superconductor for ac power transmission.

In many areas of the world, the transmission of very large blocks of power over distances of several hundred miles is increasing. As both the size of these blocks of power and the transmission distances increase, the use of dc superconducting transmission lines may be required. Nevertheless, a much more immediate problem for the electric utilities is the development of underground transmission lines to transmit large blocks of power to major urban areas from central power plants located up to 50 miles away. For this reason, I have concentrated primarily on summarizing the present status of research in ac superconducting power transmission. However, the information presented relating to the refrigerators and the cryogenic enclosures is equally applicable to superconducting dc transmission lines. In fact, these two components plus the installation costs are expected to represent over 50% of the total cost, not including right-of-way, of either dc or ac superconducting lines of comparable capacity.

CURRENT STUDIES

At the present time, there is considerable interest in superconducting power transmission in a number of countries. The major experimental studies thus far have been in the United States and Great Britain, while studies in France, Japan, Germany, and Russia have been devoted primarily to economic analyses of a number of superconducting systems. Although there is considerable amount of overlap between these various programs, they do tend in many ways to complement each other since there has been a difference in their major emphasis. For example, the study in the United States has concentrated heavily on the ac losses in superconducting niobium, the design of the cryogenic system, and a detailed economic analysis of an ac superconducting transmission line. The English studies, on the other hand, have concentrated most heavily on problems related to the electrical characteristics of a superconducting transmission line with particular emphasis being placed on problems related to line terminations and means of protecting against fault conditions.

While all current programs have a considerable way to go before the engineering development of a superconducting transmission line will be completed, the results to date are in agreement on a number of important points, the most significant of which are the following. 1. At power levels above 1 GW, the capital plus operating cost of a superconducting power transmission line will be below that of either conventional or resistive cryogenic lines.
2. Although additional engineering development is needed, there are no known fundamental obstacles which would preclude the development of superconducting transmission lines. 3. Niobium is the best choice of a superconductor for an ac superconducting transmission line.

The programs with which I am the most familiar and which, I believe, have included the most complete economic analyses are the programs carried out in the United States and Great Britain. In the United States, the largest program has been that of the Union Carbide Corporation, Linde Division, which was funded by the Edison Electric Institute while the work in Great Britain was conducted by BICC (British Insulated Callender's Cables Ltd.) on behalf of the CEGB (Central Electric Generating Board). Since there are more similarities than significant differences between all of the programs on superconducting power transmission, I will not attempt to review and compare all of these programs. Instead, I will first discuss the development and present status of the Union Carbide work. Following this, I shall review the present status of the BICC program on superconducting power transmission lines.

UNITED STATES

The most advanced program on underground superconducting power transmission in the United States is that of the Union Carbide Corporation. This program had its start in 1963 when we first began to investigate the superconducting properties of niobium prepared by an electroplating process developed by Mellors and Senderoff.[17] Measurements of both the ac losses and their relation to dc magnetization curves for this material, by Beall and Meyerhoff,[18] showed that the ac losses were low enough to make practical the use of this material in an underground ac superconducting power transmission line. In fact, Freeman,[19] in 1966, using some preliminary results of this study of ac loss in electroplated niobium wires, reported that the measured losses were only about 10% of those expected for a good cryogenic envelope of power cable dimensions. On the basis of these results, a program funded by the Edison Electric Institute was instituted in April 1968. This program consisted of two principal phases. The first was an experimental program to measure the magnetic field and temperature dependence of the ac losses in tubular niobium conductors carrying electric currents comparable to those envisioned for a superconducting power line. The second phase of the program was an economic analysis of a conceptual cable design based on the results of the ac loss measurements. This work, the details of which have been reported elsewhere,[20-24] is summarized below.

A photograph of the experimental set-up used for the ac loss measurements is shown in Figure 3. The dewar was 7 m. long and had an ID of 40 cm. One end was removable to provide horizontal access. A 40 cm. ID riser at the removable end was used for the installation of the 10,000 A current leads and a cryogen fill and withdrawal tube assembly; a second smaller riser at the opposite end was used for the electrical instrumentation. For the measurement of the ac power losses at temperatures above 4.2°K, the liquid helium

Figure 3. The ac superconducting power line test installation at Union Carbide. The transfer of liquid helium from the container on the left is being started.

---

temperature was controlled by maintaining a fixed pressure within the dewar appropriate for the temperature desired.

The ac loss measurements were made at 60 Hz on 6 m. lengths of 1 and 3 cm. diameter niobium tubes used as the inner conductor of a coaxial line shorted at one end. In all cases, the diameter of the outer conductor was nine times the diameter of the inner conductor. Two different 1 cm. diameter niobium tubes were tested. One of these was fabricated by electroplating, using the process developed by Mellors and Senderoff,[17] 0.005 cm. of niobium onto the outside of a 1 cm. OD, 0.85 cm. ID copper tube. A second 1 cm. OD conductor as well as the 3 cm. conductor and the 9 and 27 cm. diameter outer conductors were roll formed from 0.063 cm. thick annealed niobium sheet and longitudinally seam welded.

SUPERCONDUCTING POWER TRANSMISSION

In Figure 4, measured values of the 60 Hz power losses are plotted as a function of the peak value of the surface magnetic field. These data were found to fit the previously observed

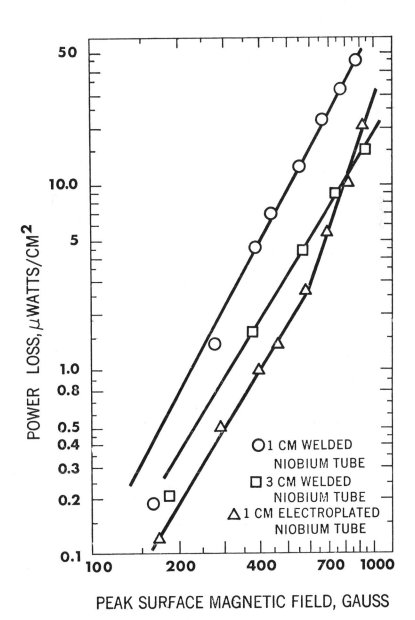

Figure 4. Power loss per unit surface area versus peak surface magnetic field for the 1 cm. electroplated and 1 and 3 cm. welded conductors

relation[18]

$$P = fA\, E_{c1} \left(\frac{H}{H_{c1}}\right)^n \quad (1)$$

where f is the frequency, A is the surface area, E_{c1} is the energy loss per unit surface area per cycle at the lower critical magnetic field, H_{c1}, and H is the peak value of the magnetic field at the surface of the superconductor. The exponent n equals 3 for $H \leq H_{c1}$ and $n \geq 4$ for $H \geq H_{c1}$.

Operation of a superconducting power transmission line with pressurized helium at temperatures above the normal boiling point of 4.2°K is preferred for a number of reasons. For example, this simultaneously avoids the difficulties associated with two-phase flow and provides for sufficient pressure drop to pump the helium through the line. Measurements made to determine the effect of temperature on the ac losses showed a linear increase of these losses of ∼ 25% as the test temperature was increased from 4.2 to 5.0°K. While these results show, as would be expected, that the ac losses increase with increasing temperature, this increase is not large enough to overcome the advantages of operating a superconducting power line with pressurized helium.

The results discussed thus far were obtained on concentric conductors in which both the magnetic field and current density are uniform about the circumference of the conductor. The three most commonly considered configurations for a three-phase superconducting power line are shown schematically in Figure 5. Two of these configurations, a and c, are symmetrical and the magnetic field and current density do not vary about the circumference of the conductors and shields. For these two configurations, the ac power losses can be calculated from (1). This is not true, however, for configuration b in which the magnetic field and current density are not constant about the circumference of the conductors and shield. An analysis of the configuration[20,21] illustrated in Figure 6 showed that for the case where the peak magnetic field at the surface of the conductor is less than H_{c1}, the power loss is given by

$$P = \frac{3R^2 + 2R + 3}{8R}\left[fAE_{c1}\left(\frac{H}{H_{c1}}\right)^3\right] \quad (2)$$

where the expression within the brackets is just the power loss for the concentric case given by (1) with $H = \sqrt{2}I/5r_1$ where I is the rms value of the total current in the inner conductor. The quantity R is the ratio of the magnetic field at the surface of the inner conductor at the point $\alpha = \pi$ to the value of the surface magnetic field at the point $\alpha = 0$ and is related to the conductor radius, r_1, the shield radius, r_2, and the displacement S, through the relationship

SUPERCONDUCTING POWER TRANSMISSION

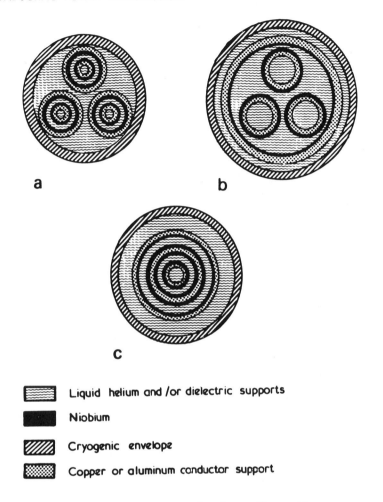

Figure 5. Schematic (not to scale) of configurations proposed for three-phase ac superconducting power lines:
    a. Triangular phase coaxial conductor and shield
    b. Triangular phase configuration single shield
    c. Concentric phase configuration

$$S^2 + S\left[2r_1 \frac{(R+1)}{(R-1)}\right] - (r_2^2 - r_1^2) = 0 \qquad (3)$$

In order to confirm the validity of (2) and (3), measurements were made with the axis of the 1 cm. electroplated conductor displaced 3.2 cm. from the axis of the 9 cm. outer conductor. For this case, R = 2 and (2) predicts that at any current, providing the peak field

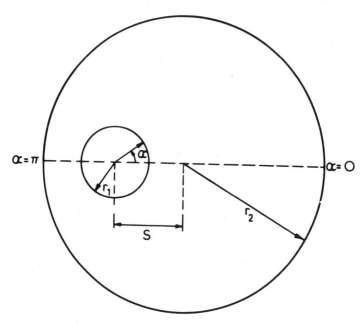

Figure 6. Schematic end view of eccentric coaxial line showing voltage probe locations and defining the parameters r_1, r_2, S and α

---

is less than H_{c1}, the ac losses will be 19/16 times greater in this eccentric configuration than would be expected for a concentric configuration. Although for practical values of S, this increase in losses is relatively small, it is important to point out that for a conductor of given radius the peak current that may be transported without exceeding H_{c1} is lower for the eccentric configuration than for the concentric configuration. Thus, while one of the simplest configurations is that illustrated in Figure 5b, the inner conductors would have to be larger than in either of the other configurations illustrated if the line current is to be the same.

The second phase of the EEI program included a conceptual design and economic analysis of a three-phase superconducting ac power line. Although several configurations were considered, a complete economic analysis was carried out only for the triangular coaxial configuration illustrated in Figure 5a. For this reason, only those design equations applicable to this configuration have been included in the following discussion.

For conventional high pressure oil filled (h p o f) paper cable, the withstand voltage, V* (kV), may be approximated by

$$V^* = 2.9V_\phi + 200 \tag{4}$$

where V_ϕ is the phase to phase voltage in kV. If an ac superconducting cable is to meet this same criterion, it has been shown[20] that the shield radius, r_s (cm.), is given by the relation

$$r_s = \frac{\sqrt{2P}}{5\sqrt{3V_\phi H}} + \frac{2.9V_\phi + 200}{E_b} \tag{5}$$

where V is the phase to phase voltage (kV), P is the line rating (kVA), H is the peak value of the magnetic field at the surface of the conductors (G), and E_b is the maximum value of the average electric field (kV/cm.) in the gap between a conductor and its shield. By minimizing this expression for r_s, with respect to V_ϕ, one can show that the optimum values of r_s, P, conductor current, I(A), and conductor radius, r_c (cm.), may be expressed in terms of V_ϕ, E_b, and H by the following equations.

$$r_s = \frac{5.8V_\phi + 200}{E_b} \tag{6}$$

$$P = \frac{17.8V_\phi H}{E_b} \tag{7}$$

$$I = \frac{10.3V_\phi H}{E_b} \tag{8}$$

$$r_c = \frac{2.9V_\phi}{E_b} \tag{9}$$

The inside radius, R_i(cm.), of the cryogenic enclosure, assuming a 10% clearance about the shields, is given by

$$R_i = 2.25r_s \tag{10}$$

Using equations 6-10, discussed above, the parameters given in Table 1 were calculated assuming that the maximum value of the magnetic field at the surface of the inner conductors would be 1,000 G, and the maximum value of the average electric field would be 200 kV/cm. Use of this value of E_b leads to a design where the peak electric fields at the surfaces of the inner conductors for the 69, 138, 230 and 345 kV lines are respectively 52, 62, 68 and 71 kV/cm. at the rated voltages. These electric stresses are well below the breakdown electric field for helium reported by Goldschwartz and Blaisse,[25] Meats,[26] Blank and Edwards,[27] and Mathes.[28] While all of the breakdown measurements made on helium to date were with gap length of a few millimeters or less, even the most pessimistic extrapolation of the available data still gives breakdown electric fields well in excess of 100 kV/cm. for gaps of

Table 1. Superconducting cable system design parameters

Line voltage, kV	69	138	230	345
Peak magnetic field, G	1,000	1,000	1,000	1,000
Dielectric strength, kV cm.$^{-1}$	200	200	200	200
Withstand voltage, kV	400	600	838	1,200
Power, mVA	423	1,690	4,710	10,590
Current, kA	3.55	7.10	11.80	17.75
Conductor diameter, cm.	2.0	4.0	6.7	10.0
Shield diameter, cm.	5.9	10.0	15.3	22.0
Cryogenic envelope, ID, cm.	13.4	22.6	34.5	49.8
Insulator thickness, cm.	5.1	5.1	5.1	5.1
Cryogenic envelope, OD, cm.	24.3	33.6	46.7	62.8
Total expense, $ per mVA mile^{-1}	1,720	601	307	201

	$ per mile	%	$ per mile	%	$ per mile	%	$ per mile	%
Conductors and shields	70,100	9.6	147,000	14.6	290,000	20.1	558,000	26.3
Cryogenic enclosure	177,000	24.3	260,000	25.6	369,000	25.6	503,300	23.7
Helium	13,300	1.9	42,500	4.2	106,900	7.4	227,000	10.7
Terminals (5-mile spacing)	30,000	4.1	40,000	3.9	50,000	3.5	60,000	2.8
Refrigeration (5-mile spacing)	121,000	16.6	149,000	14.7	168,000	11.6	177,000	8.3
Installation	234,000	32.2	265,000	26.1	317,000	21.9	420,000	19.8
Annual capitalized expenses	82,300	11.3	111,200	10.9	143,000	9.9	178,000	8.4

several centimeters. For the withstand voltages given in Table 1, these lines have peak electric fields of about 300 kV/cm. at the surface of the inner conductor. While there is essentially no data available on the breakdown electric field helium for impulse voltages, one would expect it to be higher than the reported ac and/or dc breakdown fields. It is also possible that the withstand voltage of these cables does not need to be as large as assumed since it is possible to limit surge voltages by means external to the cable. In any case, should helium prove to be incapable of supporting electric fields of the magnitude assumed there are several alternatives which have been discussed by Meyerhoff.[29] The results of Meyerhoff's analysis presented in Figure 7, shows that the cost of a superconducting power transmission line would increase by only 15% if helium were used in a design where E_b was limited to 100 kV/cm. and could even decrease somewhat if a solid dielectric having a low enough cost and loss tangent was used.

Using the measured values of the ac losses discussed above in conjunction with the design parameters given in Table 1, the capital and capitalized operating costs for ac power lines operating at 69, 138, 230, and 345 kV were estimated. The results of this economic analysis,[20] which are summarized in Table 1 and plotted in Figure 8, are based on 1969 costs and technology with no provision for technological advances which could decrease the estimated costs. For comparison, the results of a number of other economic analyses of both superconducting and resistive cryogenic power transmission lines have also been included in Figure 8.

A schematic representation of a superconducting power line installation is shown in Figure 9.

GREAT BRITAIN

In the early part of 1966, a study was initiated by BICC (British Insulated Callender's Cables Ltd.) on behalf of the Research Department of the CEGB (Central Electric Generating Board), to design a 33 kV three-phase superconducting cable having a power rating of 750 mVA as well as the design of a dc cable with a similar power rating. The initial results of this study, which were reported by Edwards and Slaughter[30] and by Rodgers and Edwards,[31] showed that the capital plus operating cost of a 33 kV superconducting line with a rating of 750 mVA would be about $1,000 per mVA mile. More recently, Edwards[32] reported that the capital cost of a 33 kV line with a rating of 4,000 mVA would be $360 per mVA mile. These results[30-32] were sufficiently encouraging to justify additional work.

The most recent results of this program, which at present appears to be oriented primarily toward superconducting ac power

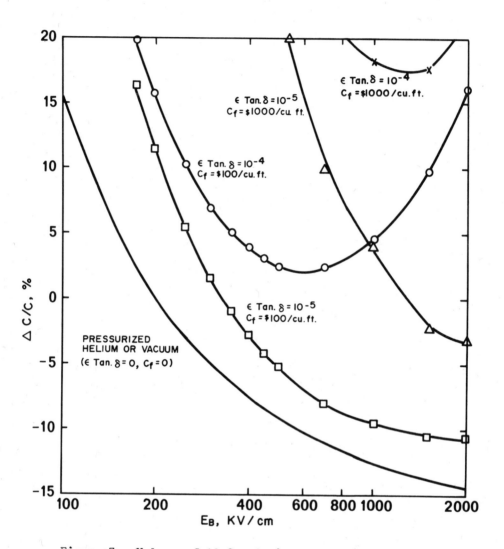

Figure 7. Values of $\Delta C/C$ in % as a function of the average electric stress, E_b, at the specified withstand voltage, V^*, where C is the original cost of the 138 kV cable employing a pressurized helium dielectric system designed for an average electric stress of 200 kV/cm. described in Table 1 (Reference 29)

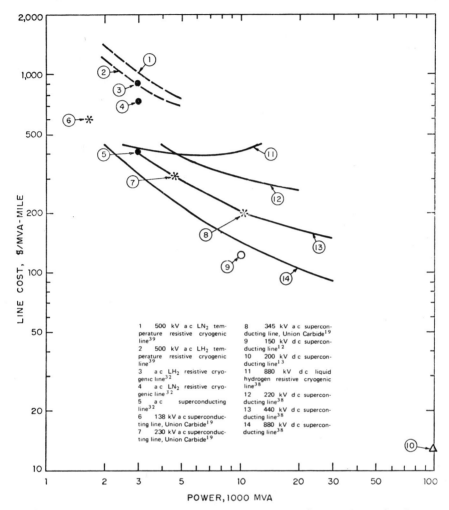

Figure 8. Cost of proposed superconducting and resistive cryogenic power transmission lines

transmission, have been reported by Rodgers et al,[33] who described work done with a laboratory scale superconducting link, Rodgers,[34] who compared resistive cryogenic and superconducting power cables, Cairns et al,[35] who discussed electrical considerations for an ac superconducting cable, and Taylor,[36] who considered the use of a layer of a type II superconductor between the niobium and the normal metal support to increase the fault current capacity of superconducting lines.

The superconducting link tested by BICC was designed to meet

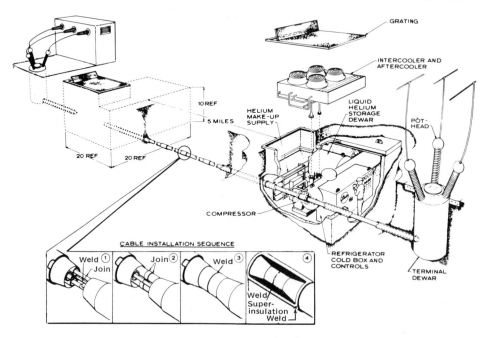

Figure 9. Illustration showing the installation of an underground superconducting ac power line proposed by Union Carbide

the following requirements.[33] "1. To provide a test-bed for measurements of ac losses on possible conductor systems of full-scale dimensions. 2. To provide practical design and operating experience of a pilot scale cryogenic system operating at liquid helium temperature. 3. To provide a laboratory demonstration of the current carrying capability of an ac superconducting system." An overall view of the superconducting link installed in the BICC Laboratory is shown in Figure 10. This superconducting link consisted of a coaxial pair of niobium conductors 2.7 m. long, having diameters of 3.5 and 5.5 cm. Difficulties encountered with the joints limited the critical current of this link to 2,080 A, considerably below the critical current of the conductors which was estimated to be about 13,000 A. Nevertheless, the current density in the niobium was about 2×10^4 A cm.$^{-2}$, which is about 200 times greater than the current density in the copper conductor in a conventional oil-filled cable.

Although the BICC-CEGB and the Union Carbide-EEI programs are similar in many respects, there are several significant differences between both the experimental programs and some aspects of the proposed superconducting transmission lines. The more significant of these differences are discussed below. First, the most

Figure 10. The superconducting link at BICC's Research and Engineering Division, Wood Lane, London. Cold nitrogen gas, from the container on the left, is being used to pre-cool the conductor system to liquid nitrogen temperature.

---

interesting difference was the use in the BICC test line of 400:1 turns ratio superconducting transformers at the ends of the test conductors. A schematic view of a transformer termination is given in Figure 11. Because of the relatively high currents generally proposed for superconducting power lines, the use of transformers to couple into and out of superconducting power lines have been suggested frequently. One can hope that the continuation of the BICC studies will provide a firm basis for evaluating the viability of this coupling method. At the present time, however, there is not enough information available to decide on either the technical or economic aspects of transformer coupling to superconducting power transmission lines. Second, because of the use of a helium boil-off method of loss measurements, the BICC cryogenic system was considerably more complex than the Union Carbide system which employed an electrical method for measuring losses. Third, the cryogenic system discussed by Rodgers[34] for the use with a

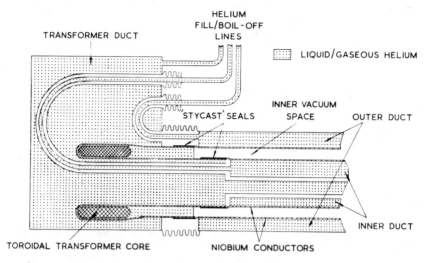

Figure 11. Schematic view of the transformer termination of the BICC superconducting link

superconducting transmission differs from that proposed by Union Carbide in that Rodgers proposes the use of a liquid nitrogen cooled shield in the cryogenic envelope while the Union Carbide cryogenic envelope uses multishields and does not require a liquid nitrogen shield. Fourth, while both BICC and Union Carbide agree on the use of niobium as the preferred superconductor for ac power transmission, BICC has proposed the use of aluminum as the conductor support while Union Carbide has proposed the use of a copper support. The choice between these two materials is very complex and depends not only on the material costs but on the cost associated with both the fabrication and joining of the conductors. These costs are not known with sufficient accuracy to make a decision as to which support material will give rise to the lower cost conductor system.

In Table 2, the results of an economic comparison of superconducting and resistive cryogenic power lines reported by Rodgers[34] are summarized. For this comparison, Rodgers considered 3,000 mVA lines operating at 132 kV, having a configuration similar to that illustrated in Figure 3a. These results are plotted in Figure 8. Although there are some significant differences between the BICC and Union Carbide proposals in both the costs assumed for various components and in the proposed designs, the total capital plus operating costs are comparable. In addition, these two studies both agree that the superconducting transmission line has lower

Table 2. Capital Costs of 3,000 mVA, 132 kV Cryogenic Cables (reference 31)

Cable type	Resistive cryogenic		Super-conducting
Coolant	Liquid nitrogen %	Liquid hydrogen %	Liquid helium %
Conductors and duct	29	10	17
Dielectric	6	3	3
Cryogenic envelope	18	9	17
LHe filling	--	--	24
LH$_2$ filling	--	17	--
LN filling	2	--	2
Duplicate refrigerators	18	41	19
Contraction joints	3	2	2
Factory assembly and inspection	6	4	4
Buildings and plant	3	2	2
Open country installation	12	9	8
Stop joints	2	2	1
Instrumentation	1	1	1
Capital cost ($ per mVA mile)	471	614	274
10 year running cost ($ per mVA mile)	282	297	142
Total ($ per MVA mile)	753	911	416

capital plus operating cost than resistive cryogenic lines operating at either liquid nitrogen or liquid hydrogen temperatures.

MAJOR PROBLEM AREAS

Although there has been a considerable amount of progress in the development of superconducting power lines since McFee's[10] initial discussion in 1962, a number of problem areas still remain. Six of the most significant of these problems are discussed below, though not necessarily in order of importance.

Line Termination

The power levels proposed for underground superconducting power lines greatly exceed the largest underground lines now in operation.

Thus, even in the absence of the difficulties associated with thermal gradient from the cryogenic environment to ambient conditions, a considerable amount of development will be necessary to develop terminations having adequate power capability.

Dielectrics

Both pressurized helium and vacuum have been considered as candidates for the dielectric system. Although pressurized helium has a sufficiently high dielectric strength to permit its use in a superconducting transmission line, additional measurements need to be made at higher voltages to determine more definitely the value of the dielectric strength to employ in the analysis and design of superconducting transmission lines. The final choice between pressurized helium or vacuum will depend to a great extent on the maximum withstand voltage for which the line is designed. In either case, however, a solid dielectric will have to be used to provide the necessary mechanical support for the conductors. So far, there is no general agreement as to the solid dielectric which best meets the requirements as to cost, ease of fabrication and installation, low loss tangent, and suitable thermal and mechanical properties.

Conductor Stabilization

The superconductor in a power line must be supported by a material capable of carrying fault currents which exceed the critical current of the superconductor. The most commonly suggested ways to accommodate the fault currents are to use a copper or aluminum backing or to use a layer of a hard type II superconductor, such as NbTi, between the niobium conductor and the normal metal conductor support. The choice between these alternatives depends in part on the magnitude of the fault current and the time required to clear the fault. The values of these two parameters have not as yet been clearly defined in terms of the requirements for underground power lines with several thousand mVA capacities.

Thermal Contraction

A suitable method for accommodating the differential thermal contraction between the conductors and the cryogenic envelope when the line is cooled must be found. Although there have been a number of suggestions as to how the differential contraction can be accommodated, at this time there is still no general agreement as to a single optimum solution.

Joints

Regardless of how a superconducting power line is designed, it will be necessary to make some joints between conductor sections. These joints must be capable of carrying the normal rated line current and perhaps even a fault current without contributing an excessive refrigeration load. The severity of this problem is, of course, dependent on the number of joints which will be required. While the joints are certainly not an insurmountable problem, this is an area in which additional research and development is required.

Cryogenics

Last and certainly not least is the development of cryogenic envelopes and refrigerators which will have the long term reliability and ease of maintenance required for a system to be used by the electric utilities.

SUMMARY

The studies carried out to date clearly demonstrate that underground superconducting power lines promise a solution to the power transmission problems predicted for the not too distant future. Although the development of superconducting power lines may take somewhat more time than the development of resistive cryogenic lines, the superconducting lines, as illustrated by the data presented in Figure 8, are clearly less costly and will therefore be of greater importance in the long run. For example, at power levels above 2,000 mVA, the capital plus operating cost of a superconducting transmission line is less than 50% of the costs of either resistive cryogenic or conventional underground lines of equivalent rating.

Although there are a number of studies of superconducting transmission lines under way throughout the world which when taken together cover most of the major problem areas, there are still a few areas in which a considerable amount of additional engineering development will be required before superconducting transmission lines are ready for service. A 12-1/2-year program proposed by the Linde Division of the Union Carbide Corporation to the Edison Electric Institute for the completion of the engineering development and prototype testing of such a power line is outlined in Figure 12. Although I am not familiar with the details of the future programs envisaged by others to develop such power lines, I would guess that their time tables do not differ very greatly from the program outlined in Figure 12. Thus, it is apparent that little time is to be lost if superconducting power lines are to be ready to fulfill their destiny in the decades following 1980.

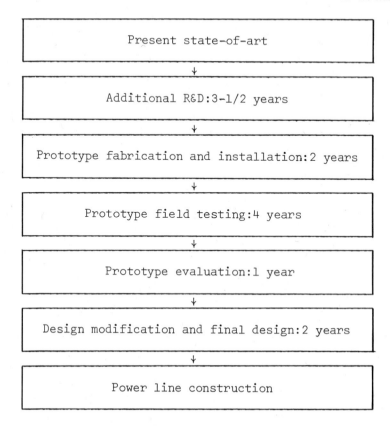

Figure 12. Outline of the program proposed by Union Carbide for the completion of the development of ac superconducting power lines.

The author is pleased to acknowledge the assistance of Mr. D.A. Haid in the preparation of this paper.

REFERENCES

*This paper is based on a recent review article by the author [Cryogenics 11, 91 (1971), published by IPC Science and Technology Press, IPC House, 32 High Street, Guildford, Surrey, England].

1. Considerations Affecting Steam power Plant Site Selection (Office of Science and Technology, Washington, DC, 1968)
2. Kusko, A., IEEE Spectrum 5, 75 (1968)
3. Boesenberg, E.H. and Zanona, A., 33rd Annual Meeting, American Power Conference, Chicago, (April 1971)
4. Underground Power Transmission, A Report to the Federal Power Commission by the Commission's Advisory Committee on Underground Transmission (Washington, DC, 1966)

5. Cullen, A.L., Endeavor 29, 55 (1970)
6. Paul, H., Electronics and Power 16, 171 (1970)
7. Graneau, P., IEEE Trans Power Apparatus and Systems PAS-89, 1 (1970)
8. Minnich, S.H., and Fox, G.R., Cryogenics 9, 165 (1969)
9. Pastuhov, A., and Ruccia, F., Conference on Low Temperatures and Electric Power, IIR, London, 1969
10. McFee, R., Electrical Engineering 81, 122 (1962)
11. Atherton, D., Paper No. 3121, IEEE Canadian Electronics Conference, Toronto, Canada, 1963
12. Klaudy, P.A., Advances in Cryogenic Engineering, Vol. 11, 684 (Plenum Press, 1966)
13. Gauster, W.F., Freeman, D.C., and Long, H.M., World Power Conference, Lausanne, Switzerland, 1964
14. Garwin, R.L., and Matisoo, J. Proc IEEE 55, 538 (1967)
15. Voigt, H., Conference on Low Temperatures and Electric Power, International Institute of Refrigeration, London, 1969
16. Delile, G., Conference on Low Temperatures and Electric Power, International Institute of Refrigeration, London, 1969
17. Mellors, G.W., and Senderoff, S., J. Electrochem.Soc. 112, 266 (1965)
18. Beall, W.T., and Meyerhoff, R.W., J. Appl. Phys. 40, 2052 (1969)
19. Freeman, D.C., Applied Superconductivity Conference, Brookhaven, New York, 1969
20. Long, H.M., Beall, W.T., Eigenbrod, L.K., Meyerhoff, R.W., and Notaro, J., Edison Electric Institute Project RP78-7, Final Report, October 1969
21. Meyerhoff, R.W., and Beall, W.T., J. Appl. Phys. 42, (1971)
22. Eigenbrod, L.K., Long, H.M., and Notaro, J., ASME Winter Annual Meeting, Los Angeles, California, 1969
23. Eigenbrod, L.K., Long, H.M., and Notaro, J., Paper No. 70TP168-PWR, IEEE Winter Power Meeting, New York, 1969
24. Long, H.M., and Notaro, J., J. Appl. Phys. 42, (1971)
25. Goldschuartz, J.M., and Blaisse, B.S., Conference on Low Temperatures and Electric Power, International Institute of Refrigeration, London, 1969
26. Meats, R.J., International Conference on Gas Discharges, September, 1970
27. Blank, C. and Edwards, M.H., Phys. Rev. 119, 50 (1960)
28. Mathes, K.N., IEEE Trans. on Electrical Insulation, EI2, 24, (1967)
29. Meyerhoff, R.W., To be presented at the Electrical Insulation Conference, Chicago, September, 1971
30. Edwards, D.R., and Slaughter, R.J., Electrical Times (3 August 1967)
31. Rodgers, E.C., and Edwards, D.R. Elect. Rev., 348 (8 September 1967)

32. Edwards, D.R., The Electrical and Electronics Technician Engineer (1968)
33. Rodgers, E.C., Cave, E.C., and Grigsby, R., Conference on Low Temperatures and Electric Power, International Institute of Refrigeration, London, 1969
34. Rodgers, E.C., Conference on Low Temperatures and Electric Power, International Institute of Refrigeration, London, 1969
35. Cairns, D.N.H., Minors, R.H., Norris, W.T., and Swift, D.A., Conference on Low Temperatures and Electric Power, International Institute of Refrigeration, London, 1969
36. Taylor, M.T., Conference on Low Temperatures and Electric Power, International Institute of Refrigeration, London, 1969

APPLICATION OF SUPERCONDUCTIVITY IN THERMONUCLEAR FUSION RESEARCH [+]

A. P. Martinelli

Max-Planck-Institut für Plasmaphysik

8046 Garching bei München, West Germany

INTRODUCTION

This paper describes the application of superconductivity to controlled fusion research at Max-Planck-Institute of Plasma Physics in Garching, i.e. mainly in the quadrupole (Wendelstein 6) and stellarator (Wendelstein 7) experiments, which are respectively under construction and at the advanced planning stage.

First, however, it is convenient to examine the recent technological progress in the field of superconducting materials and then to examine briefly the most important plasma experiments in Western Europe and in the United States which use superconducting magnets.

The paper finishes by outlining the part that will be played by superconducting magnets in possible future thermonuclear reactors.

RECENT DEVELOPMENTS OF HIGH FIELD SUPERCONDUCTORS

The properties of a superconducting material are described by the three parameters, critical magnetic field H_c, critical current density J_c, and critical temperature T_c. These parameters define a surface in a three-dimensional representation H, J, T which divides the space H, J, T into a region where the material is superconducting and another where the material is normal.

Critical Temperature T_c

A high T_c value is desirable both because the efficiency of refrigerators (ideally the efficiency of the Carnot cycle) rapidly decreases as the temperature necessary to keep the material superconducting drops, and because a material having a higher T_c is less sensitive to thermal disturbances, the other parameters such as operating temperature, critical current, stabilization etc. being the same. This material will undergo normal transition less easily than a material having a lower T_c since $I_c(T)$ decreases about linearly with increasing T.

The ternary compound $Nb_3(Al_{0.8}Ge_{0.2})$ is the material having the highest critical temperature of $T_c = 20.7°K$. The compound Nb_3Sn has $T_c = 18°K$, and the alloy Nb-Ti has $T_c = 10$ °K. A T_c of about 25°K is considered to be the upper limit for materials such as those mentioned above, in which the superconducting state is induced by phonons. Other mechanisms susceptible of generating superconductivity at higher temperatures have been suggested (e.g. organic superconductors), but materials for technological applications do not appear probable in the near future.

Critical Magnetic Field

Recent measurements show that the compound $Nb(Al_{0.8}Ge_{0.2})$ has a critical field $H_c = 410$ kOe, Nb_3Al[1] has $H_c = 295$ kOe, and V_3Ga[2] has $H_c = 220$ kOe.

Although the first two materials are not commercially available, the recent development of high field materials is promising for the future realization of magnets capable of producing fields higher than those obtained up to now by using Nb_3Sn ($H_c = 210$ kOe).

The third material, V_3Ga ($T_c = 14.5°K$), which recently appeared on the market at a price competitive with that of Nb_3Sn would afford, according to recent measurements[2], a considerable advantage over Nb_3Sn in the construction of magnets generating fields greater than 120 kOe. This can be seen by considering the product $J_c \cdot B_c = F_v$. It has the dimensions of a force per unit volume and corresponds physically to the strength of the pinning forces in the superconductor volume. The product F_v has a maximum at about 175 kOe

(0.8 H_c) for V_3Ga (F_{vmax} = 1 to 2 x 10^3 kg/cm^3), while for Nb_3Sn it has about the same maximum value F_{vmax} at 120 kOe and then drops rapidly. Therefore, for fields higher than 120 kOe a magnet using Nb_3Sn requires a greater volume of superconducting material than an equivalent magnet made of V_3Ga, so that using V_3Ga is more economical than Nb_3Sn, assuming the same specific cost for both materials.

Critical Current Density J_c and Stabilization

The high values of the critical current density J_c (> 10^3 A/mm^2) that have been reached in short samples of high field superconductors since their discovery are considerably reduced (degradation) once the superconductor is wound in a coil.

In order to limit the degradation of the superconductor, this is stabilized by, for instance, adding a material of low electrical and thermal resistivity, e.g. copper. Although the cross section of the composite, copper plus superconductor, is larger than that of the superconductor alone, the current density obtained in the composite is higher than that reached in the degraded superconductor. Therefore the current density value is a function of the type and extent of stabilization to which the superconductor is subjected.

Cryogenic Stabilization.
In the case of "cryogenic" stabilization a material of high electrical and thermal conductivity is in close contact with the superconductor so as to take over part or all of the current which the superconductor, partially or totally turned normal owing to a thermal or magnetic instability, cannot momentarily transport. Let us suppose that the current momentarily flowing in the stabilizing material of resistivity ρ has a density J such that the heat $q = \rho J^2$ (watt/cm^3) is generated. Let us suppose further that the total heat $Q = qV$ generated in the volume V of the stabilizing material is smaller than the maximum power which can be transferred in the "nucleate boiling" regime from the surface of the composite to the surrounding liquid helium. The composite can then revert to thermal equilibrium and the current will again flow in the superconductor. The composite is thus "stable". The critical value of the power transferred from the surface unity of the composite to the liquid helium in

the nucleate boiling regime is approximately q_c = 0.5 watt/cm^2, beyond which the heat transfer process switches from nucleate boiling to "film boiling". Correspondingly the temperature differences ΔT between composite surface S and liquid helium passes from a few tenths of degree in the nucleate boiling regime to a few degrees in the film boiling regime [3].

Therefore, if the current density J is such that $Q > Q_c = qS$, the superconductor will turn completely normal and will revert to superconductivity only after the current has been noticeably reduced or reduced to zero. In this case the composite is "unstable".

A high current density of the composite is generally required, and for reasons of economy the current density in the superconductor should be close to the critical value. On the other hand, the copper-to-superconductor ratio of the composite as well as the composite cooling environment determine the stability behavior as described above.

Bearing the above conditions in mind, one can consider a cryogenically stabilized copper-superconductor composite as being in a state of stability, limited stability, or instability, if the total current density is approximately J < 80 A/mm^2, J = 100 - 200 A/mm^2, and J > 300 A/mm^2 respectively [4].

<u>Adiabatic Stabilization.</u> Under any operating conditions, apart from when the superconducting material carries its maximum current, the current distribution in the superconductor consists of a transport current and a local current ("magnetization" or "shielding" current). The latter can decay in a short time (a few tens of /usec) owing to a thermal or magnetic disturbance, and the associated energy is liberated thermally (flux jumps). The related temperature rise of the superconductor can ultimately lead to a quench.

Theoretical and experimental work shows that such shielding currents can be substantially reduced or eliminated by decreasing the diameter of the individual superconducting strands embedded in the matrix (copper or cupro-nickel) to a value of about 20 /u or smaller and by transposing them [5]. The filamentary structure thus obtained is said to be adiabatically, intrinsically or enthalpically stable. This stabilization method probably represents the most significant contribution to superconducting technology in the last two or three years.

By this technique the copper constituting the matrix of the filamentary composite can be reduced, thereby achieving copper-to-superconductor ratios of 1 to 1 and total current densities in excess of 300 A/mm^2. Furthermore, no special cooling arrangements such as liquid helium cooling channels are needed for the magnets, thus allowing more compact mechanical design and higher overall current densities. The coils can be potted with epoxy and give the same performance when operated in liquid or gaseous helium, since they are adiabatically stable and not directly affected by the cooling environment.

Dynamic Stabilization. A third stabilization method is referred to as dynamic stabilization, which allows for the fact that the thermal diffusivity of the superconducting material (D_{th} = K/c, where K is the thermal conductivity, and c the specific heat) is much smaller (factor 10^3 for Nb-Ti) than that of copper, and that the thermal diffusivity is much smaller than the magnetic diffusivity $D_{mag.}$ = $\rho 10^9/4\pi$ (the opposite is true of copper). Consequently, the heat generated in the superconductor during penetration of the magnetic field following a sudden variation of the field itself (flux jump) does not get removed at the same rate at which it is produced, and therefore causes dangerous temperature rises.(4)

This method consists in embedding the superconductor in a metal matrix (e.g. copper), the purpose here being to transfer rapidly the generated heat from the superconductor to the helium bath and to shield, at least partially, the superconductor from variable magnetic fields, either externally superposed on the winding or internally caused by flux jumps taking place in the superconductor located in the more or less immediate neighbourhood, i.e. in the same layer or in contiguous layers.

This method also covers the known technique of interleaving thin copper sheets between two or more contiguous layers of a winding. The sheets are thermally and electrically insulated from the layers and are intended both to shield the superconducting layer from the variable magnetic fields and directly transfer the heat, generated by these fields, to the helium bath, with which the sheets are in good contact.

Actually, there is a certain overlapping of the three stabilization methods, which are described above as separate, and in practice one of them will be predominant.

APPLICATION TO PLASMA PHYSICS

Owing to the interaction of electrically charged particles with a magnetic field, the latter is ideal for confining a plasma, which is composed of charged particles (ions and electrons) in which a gas at a high or very high temperature ($> 10^4$ °K) is dissociated. In fact, if the magnetic field is sufficiently high, the trajectory of each particle takes the form of a spiral with the axis parallel to the magnetic field, which therefore acts as a non-material container and, given the high temperature of the plasma, is incombustible.

Magnetic Configurations

If the confining magnetic field is so constituted that none of its field lines leaves the confinement volume of the plasma, the particle and energy losses of the plasma can only take place in a direction perpendicular to the magnetic field and are limited. These conditions are met in "closed" magnetic configurations, e.g. in toroidal machines such as the "Stellarator" and "Tokamak".

In "open" configurations, on the contrary, serious plasma losses occur at the two ends, where the field lines leave the plasma confinement volume. One attempts to reduce these losses by considerably increasing the magnetic field value at the ends to produce "magnetic mirrors" so as to reflect the plasma.

The experimental results achieved in open configurations (mirror machines, theta pinches) have been better, as far as the temperature and density of the plasma are concerned, than in closed configurations, where, on the other hand, better results have been obtained for the confinement time.

One tends to believe now that a possible future thermonuclear reactor will preferably have a closed geometry, especially owing to the absence of the end losses present in open configurations. It should be noted, however, that the open-ended or closed configurations built up to now or still in the construction or design stage are only intended for studying the properties of the plasma, viz. temperature, density, stability and confinement etc., and do not even represent miniature prototypes of thermonuclear reactors.

While there is little doubt that a future thermonuclear reactor will utilize superconducting magnets owing to both the large dimensions and high field required, several experiments using superconducting magnets have recently been built or are being planned. The main reasons for this are as follows:

- a) The magnetic fields necessary for the experiments have increased both in volume and in magnitude. They can now be more economically and conveniently realized with superconductors than with traditional techniques. Furthermore, many experiments require the use of stationary instead of pulsed magnetic fields.

- b) High current densities are possible particularly if intrinsically stable superconductors are used. The resulting reductions in volume of the magnets allow better access for plasma production and diagnostics. A current density $J = 100 - 200$ A/mm^2 is generally required, which would correspond to a region of limited stability in the case of cryogenic stabilization.

- c) Reliability of superconducting magnets, recently shown by the satisfactory operation of a number of large magnets, e.g. the magnet for the bubble chamber at the Argonne National Laboratory, which is 4.80 m in diameter and produces 18 kOe at the center.[6]

Table I, compiled by Hancox [7], gives an up-to-date picture of the plasma experiments either in operation or under construction in the United States and Western Europe which make use of superconducting magnets and shows that most plasma laboratories are engaged on at least one important project applying superconductivity.

The Application of Superconducting Magnets at Garching

The work on superconductivity at Garching was initially restricted to the construction of a few small

Table I
Plasma physics experiments using superconducting coils

Laboratory	Type	Dimensions (cm)	Field or Current	Material
	Open Configurations			
Oak Ridge N.L. (USA)	I.M.P. Mirror and Quadrupole	35 dia 70 long	20 kG central 75 kG peak	Nb-Ti + Nb_3Sn
N.A.S.A. Lewis (USA)	Mirror	51 dia	75 kG central	Nb-Ti + Nb_3Sn
L.R.L. Livermore (USA)	Baseball II	120 dia	20 kG central 75 kG peak	Nb-Ti
	Closed Configurations			
Princeton P.P.L. (USA)	Spherator	90 ring dia	130 kA	Nb_3Sn
Princeton P.P.L. (USA)	F.M.1 Multipole a) 1 ring b) 2 rings	150 ring dia 100 " " 200 " "	375 kA 460 kA 325 kA	Nb_3Sn
L.R.L. Livermore (USA)	Levitron	80 ring dia	500 kA	Nb_3Sn + Nb-Ti
Culham (England)	Levitron	60 ring dia	500 kA	Nb-Ti
Garching (West Germany)	Wendelstein 6 Multipole	60 ring dia 120 " "	400 kA 190 kA	Nb_3Sn
Garching (West Germany)	Wendelstein 7 Stellarator	400 maj. dia 40 min. dia	40 kG mean 65 kG peak	Nb-Ti

magnets for laboratory use and to the evaluation of new commercial materials.

<u>Module system of 100 kOe.</u> A system of superconducting magnets with much larger dimensions was built during 1967 - 1968 for the purpose both of obtaining a field of 100 kOe in a 5 cm bore for material testing and of developing superconducting magnets with the dimensions required for a number of plasma experiments [8]. Particular attention was paid to developing standard types of coils for assembly into different configurations (e.g. mirror configu-

rations, cusp fields, etc.), as already done with copper coils.

Thus, three different coils of the modular type have been designed and built. These have inner diameters of 350, 170, and 50 mm and are referred to in Fig. 1 as SSp 350, SSp 170, and SSp 50. Fig.1 shows an assembly of 12 coils (four of each type concentrically mounted). This system produced a maximum field of 105 kOe.

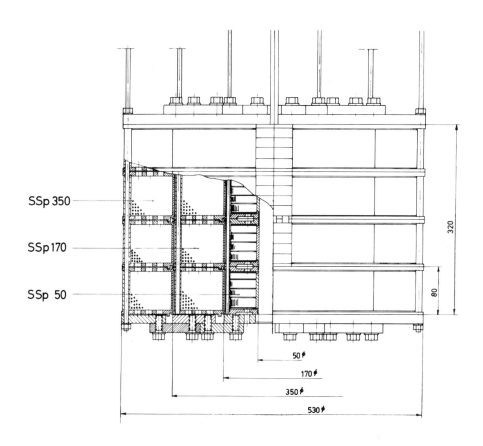

Fig.1. Assembly of twelve superconducting coils SSp 350, SSp 170, and SSp 50 (four of each) forming a 100 kOe magnet system.

The composite used for the coils SSp 350 and SSp 170 (Supercon, Atomics International) is a cable made of strands of Nb-Ti and copper twisted together and impregnated with indium. This method of manufacturing superconducting composites became obsolete soon after and was replaced by the manufacturing process used now. This consists in embedding the Nb-Ti in the copper ingot already at the beginning of the mechanical manufacturing process and then in swaging the ingot down to the final composite diameter. Excellent contact of a metallurgical type is thereby assured between copper and Nb-Ti.

The field strength and critical current density (100 - 200 A/mm^2) reached by these coils are very close to those of the short sample. This was not the case for the coils of the type SSp 50, which are wound with Nb_3Sn ribbon (General Electric, Radio Corporation of America). These show considerable degradation due to the poor stability behavior of the Nb_3Sn ribbon. This behavior is caused by the magnetization currents, which increase with increasing width of the ribbon (12,7 mm in this case) and are induced by magnetic fields in the coil radial direction, i.e. perpendicular to the ribbon surface [9].

<u>Wendelstein 6 Quadrupole (W 6)</u>. Taking a closer look at plasma experiments, one observes from Table I that four laboratories are engaged on experiments making use of magnetically supported superconducting rings.

Figure 2 represents schematically the quadrupole configuration with the two superconducting rings carrying the currents of density J_1 and J_2, the resulting magnetic field B, and the plasma distribution. It is hardly conceivable that a future thermonuclear reactor will have this configuration since the current carrying and field generating rings are entirely surrounded by the plasma. However, this configuration is of great interest for the study of plasma confinement and stability, essentially for two reasons: a) The minimum of the confining magnetic field in the region between the rings offers the plasma a particularly stable MHD equilibrium condition; b) The uncomplicated geometry of the configuration allows relatively simple theoretical treatment of the experimental results [10].

In previous similar experiments a current pulse was sent through the rings to produce the magnetic field. The rings were sustained either by thin supports or even left momentarily without supports for the short

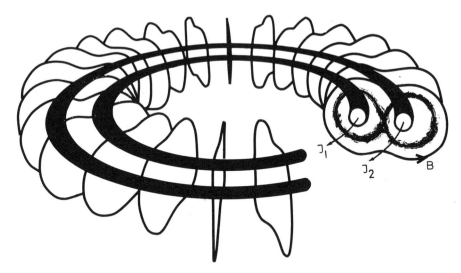

Fig.2. Toroidal quadrupole (schematic).

duration of the experiment. However, the plasma containment was negatively affected by plasma losses at the supports, which, moreover, generated electric fields in the plasma (these electrical fields in turn causing additional plasma losses). Finally, the turbulence associated with the necessarily rapid injection of the plasma in the case of the rings momentarily left without supports also had an adverse effect.

Therefore, the use of levitated superconducting rings offers an ideal solution because high and constant magnetic fields are achieved, no material supports are required, and the experiment can last minutes or hours. In this specific case the application of superconductivity allows a configuration which could not be constructed otherwise. A cost comparison between the use of superconducting or non-superconducting technique is therefore neither here nor there.

Through the magnetic supports the rings have no

mechanical or electrical contact with the environment, and the thermal connection with it is limited to radiation. The radiation losses are responsible for the gradual heating of the rings to the transition temperature T_c.

Since a long operating time of the device is desirable, once the rings have been cooled down to a temperature close to $4.2°K$, it is evident that commercially available superconducting materials having a high transition temperature such as Nb_3Sn are an obvious choice (Spherator at P.P.L., Princeton; Levitron at L.R.L., Livermore; Quadrupole W6 at I.P.P. Garching). However, despite its lower transition temperature ($10°K$), intrinsically stable Nb-Ti has been preferred to Nb_3Sn in the Culham project (Levitron) owing both to the lack of knowledge about the operation of Nb_3Sn coils in gaseous environment and to the high contents of magnetization currents necessarily present because of the Nb_3Sn tape width (not obtainable in filament form), these currents being likely to distort appreciably the magnetic field and consequently affect the containment of the plasma. Finally, the lower cost of Nb-Ti in comparison with Nb_3Sn has also been taken into consideration.

The utilization of liquid helium for the cooling and operation of the rings implies the use of cryostats around the rings and therefore a larger total volume of the rings and of the whole experiment (e.g. Spherator at Princeton). This is one of the main reasons why helium gas has been preferred at Livermore and Garching.

At Garching a series of tests have been carried out just to establish the behavior of Nb_3Sn in a gas environment. These consisted of measurements of the characteristics $(H, J, T)_c$ of short samples to establish particularly the maximum temperature the superconducting rings can reach before reverting to normality when they carry the nominal current $I_n = 100$ A; magnetization measurements at different temperatures, particularly to see at what temperature the rings can be most conveniently energized to reduce the diamagnetic currents, both to increase the magnetic stability of the superconducting windings and to reduce the field distortion; measurements on small superconducting coils carrying the same current and in the same field as the W6 rings, the temperature of which is varied to obtain an indication of the stability of the rings when their temperature increases (11).

Figure 3 shows schematically a design cross section of the quadrupole W6[12]. The two rings have a diameter of 60 and 120 cm, and are wound with 4,000 and 1,900 turns respectively. For both rings the nominal current is I_n = 100 A. As seen in Fig. 3, the winding of the small ring consists of six double pancakes, while that of the large ring has four double pancakes. Each pancake is 5 mm thick, equal to the width of the Nb_3Sn tape used (General Electric). The tape is copper stabilized, is reinforced with a stainless steel ribbon 0.025 mm thick, and has a total thickness of about 0.2 mm. The joints in the winding are obtained by overlapping the tape ends for about 10 cm and by tin soldering. Their resistance is smaller than $10^{-8}\,\Omega$. Fig. 3 shows also the rectangular cross sections of six copper coils two by two symmetric with regard to the plane AA, and located externally to the vacuum container (see Fig. 4c) which encloses the rings. Their purpose is to shape the field produced by the rings. The two remaining copper coils indicated in the lower part in Fig. 3 support the two rings by producing a field equal to 0.1 to 0.2 % of the total maximum field reached in the configuration. The small and the large ring weigh 200 and 400 kg respectively, and the magnetic field reaches a maximum of 30 kOe at the inner boundary of the small ring.

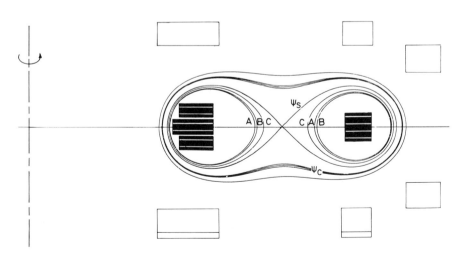

Fig. 3   Wendelstein 6 quadrupole (design scheme).

The singular field line named ψ_s in Fig. 3 is called the separatrix and has approximately an 8 shape in a quadrupole [13]. This line divides the magnetic configuration into two regions, the field lines in the one closing around one ring, and in the other around both rings. At the crossing of the separatrix the magnetic field B goes to zero; the configuration is of the type "zero-minimum B". Inside the lines ψ_c the plasma is stable (MHD stability). The three curves A, B, C, correspond two by two to three possible configurations of the plasma confinement. The innermost curves delineate the material sheaths of the rings. These sheaths are vacuum tight and contain the windings constituting the superconducting rings, the relevant mechanical structure, and the other devices described in Fig. 4a.

Figure 4 shows schematically some technical, particularly cryogenic, details of W6 [14]. In Fig. 4a the inductance L indicates the superconducting winding; the switch S closes after the generator G has energized both windings; finally, the electrical connections between generator and superconducting rings are withdrawn. In this fashion a practically permanent current flows in the rings; in the relevant time constant $\tau = L/R$, R is given practically by the resistance of the switch in the closed state (about $10^{-7} \Omega$). The switch is an electro-mechanical one, has an infinite resistance in the open state, and makes use of superconductors (Nb_3Sn ribbon) to achieve in the closed state the said low resistance contact. The switch is now being built.

It has been established that the rings are most conveniently energized as close as possible to the critical temperature T_c [11]. This is done first by superposing the magnetic field produced by the system of external coils (except those for levitating the rings) on the rings while they are at a temperature $T > T_c$. The rings should then be gradually energized while being cooled to below T_c. Care should thereby be taken to reach the nominal values of the current (100 A) and field by moving along the critical surface $(H,J.T)_c$ and finally cooling further to 4.2°K. This way of energizing should ensure the best magnetic stability of the rings and a minimum of magnetization in the windings, i.e. a minimum of field distortion.

To increase the useful experimental time, it is planned to increase the thermal capacity of the rings by incorporating a metal of high enthalpy (K in

Fig. 4a). For this purpose mercury is slightly more advantageous than lead, but the latter is preferable since it is not so toxic as mercury.

The rings are cooled by means of three vertically retractable supports (SR in Fig. 4b) spaced at 120° intervals on a circumference of diameter equal to the average diameter of the two rings. The supports have the double purpose of mechanically supporting the rings when they are not in operation and contact cooling them

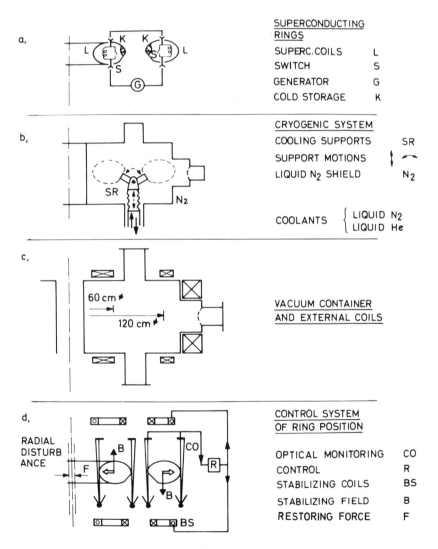

Fig.4. Wendelstein 6 quadrupole (technology).

from 300°K to about 4.5°K by using first liquid nitrogen and then liquid helium. The top part of the supports can rotate a few degrees, as indicated in Fig. 4b, so that they can readily be detached from the two rings once the latter have been cooled.

The shield N_2 in Fig. 4b is kept at liquid nitrogen temperature and has the purpose of slowing down the gradual heating of the rings due to radiation.

Figure 4d schematically illustrates a control system for the ring position, which uses photocells as sensors.

A series of measurements is now being concluded on a test coil with an inner diameter of 22 cm and made of two double pancakes wound with the same material as the W6 rings will be made of. After these have been completed the W6 rings will be wound and will be tested by the end of the year.

Wendelstein 7 Stellarator (W7). In the last years plasma confinement times close to the maximum theoretical values and approximately classical diffusion conditions have been achieved at Garching in a stellarator of relatively modest dimensions[15]. To continue these experiments and particularly to extend them to a magnetic field up to 40 kOe, a large stellarator is now under development with a major diameter of 4 m and minor diameter of 0.4 m. Since such a high field should be stationary, it is convenient to generate it with superconducting magnets.

The dimensions and cost of this machine (of the order of two million dollars for the superconducting magnets only) are considerable, but this machine can be considered as one of the essential steps prior to construction of a first "laboratory" thermonuclear reactor in which a deuterium-tritium plasma reaches the characteristics of density, temperature, and confinement time corresponding to Lawson's criterion for a thermonuclear fusion plasma [16].

Figure 5 illustrates the general design characteristics of the stellarator W7 [17].

Forty superconducting magnets are arranged along the torus 4 m in diameter and produce the toroidal field of 40 kOe in a useful circle 0.4 m in diameter. The minimum number of magnets is primarily imposed by

the maximum field ripple which can be tolerated for the
experiment. Each magnet is provided with its own cryostat.
This solution has been preferred since it allows easier
assembly and disassembly, access to the plasma,
possibility of modification of the magnet system compared
with the case where all the magnets had a common cryostat.
Furthermore, it allows a final adjustment of the magnet
system which is most important owing to the severe
mechanical tollerances imposed by the magnetic configuration of the stellarator.

Finally, the magnet with its own cryostat constitutes
a module which can be variously combined with other
units to form different magnetic configurations according

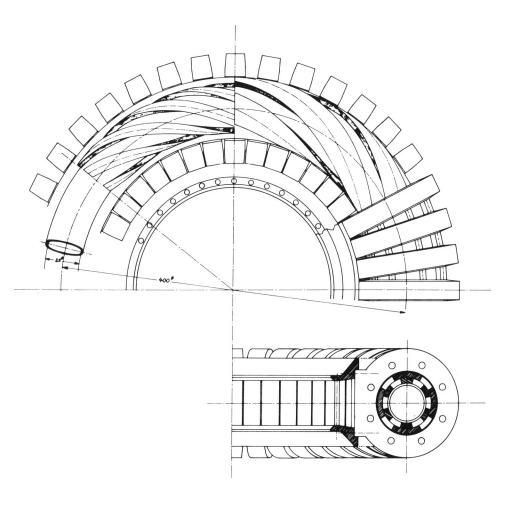

Fig. 5. Wendelstein 7 stellarator (general view).

to the concept already mentioned in relation to the system shown in Fig. 1.

Besides the advantages mentioned, this solution of having a cryostat for each magnet entails considerable complications. In fact, the magnetic forces to which each of the 40 magnets is subjected from the others have to be transferred from the superconducting winding at helium temperature, through the cryostat, to room temperature, before they are transferred to the mechanical supporting structure. For this purpose it is intended to use supports of glass fiber reinforced epoxy which offer a maximum value of the ratio σ/k between compression resistance and coefficient of thermal conductivity [18]. However, even under optimum conditions these supports cause a heat flux of about 7 watt, which represents approximately 70 % of the total heat flux in each coil (1.5 watt for the current leads, 0.5 watts for radiation losses, 1 watt for the helium transfer line, making a total of 10 watt). Therefore, for the system of 40 coils one must have a refrigerator of at least 400 W.

A second complication inherent in the concept of a separate cryostat for each coil arises from the fact that each cryostat extends in azimutal direction (along the circumference of the torus) a certain distance which is subtracted from the space available for the winding. Consequently, the current density in such a winding is bound to be considerably higher than in the case of a system of coils placed in a common cryostat. A high current density is therefore necessary in the winding (about 111 A/mm^2), which can be obtained by using intrinsically stable superconductors.

The prototype of the planned 40 coils is now under construction. The data at this stage of development are given in Table II.

While the use of intrinsically stable superconductors permits safe operation in ordinary conditions, the possibility is not excluded that a magnet might suddenly quench, e.g. owing to a thermal disturbance. In this case a protection system can be provided to extract a large part of the magnetic energy of the coil under transition so as to prevent the superconductor from being damaged owing to overheating.

Should the current in a coil suddenly drop to zero because of the destruction of part of the winding or

APPLICATION IN THERMONUCLEAR FUSION RESEARCH

Table II
Data of prototype coil for the stellarator W 7

Conductor: intrinsically stable, rectangular cross section, 3.35 x 1.42 mm, 580 Nb-Ti filaments 50 μ in diameter in copper matrix, copper/superconductor ratio 3.1/1.

Winding: 12 double pancakes

Nominal current	I_n = 700 A
Number of turns	N = 1,430
Total ampere turns	NI = 10^6 A
Inner diameter of coil	D_i = 950 mm
Outer diameter of coil	D_o = 1,130 mm
Current density in conductor	J = 147 A/mm^2
Current density in winding	J_w = 111 A/mm^2
Inductance	L = 2.9 H
Energy stored	E = 1.4 MJ
Maximum field at winding	H_w = 45 kOe
Central field	H_o = 12.2 kOe

System of 40 coils

Maximum field at winding	H_w = 65 kOe
Central field	H_o = 40 kOe
Energy stored	E_t = 57 MJ

---

interruption of the conductor, the remaining coils are pressed together with a force of about 200 tons; the force is even larger, up to 600 tons, if instead of one coil a whole sector suddenly gets deenergized. In this case a protection system causes deenergization of the other magnets to avoid destructive mechanical stress.

Figure 5 shows further two independent helical copper windings concentric with the superconducting magnet system. They are for inducing a quadrupole or hexapole field along the torus so as to cause a rotational transform of the field lines.

It is necessary for the experiment that the constant magnetic fields, approximately 5 kOe in amplitude, which are produced by the helical windings, reach the maximum in a rise time of a few msec after being switched on. During the rise time the variable fields permeate the superconducting windings, which can revert to normality, both owing to the direct effect of the increased field and to the heating caused in the composite copper-superconductor by eddy currents and hysteresis losses induced by the field variation. Calculations and experimental work show that in the specific W7 configuration the helical field rise time in the superconducting windings should be slowed down (e.g. by means of copper shields) to at least 0.2 sec.

This problem, i.e. the superposition of a variable magnetic field on a stationary field produced by superconducting magnets, is inherent in several plasma configurations and has already been treated in configurations where the variable magnetic field is pulsed, has rise times much shorter (5-10 μsec) than those considered for the W7, and returns to zero in an equally short time after reaching the maximum value $(19, 20, 21)$.

Apart from this problem, which has required and still requires particular theoretical and experimental investigation, the project W7 does not deviate from the relatively conventional superconductivity application. For this reason, and because of its dimensions and time schedule, the W7 development has been contracted to an industrial firm (Siemens Central Laboratories, Erlangen) which has enough trained personnel to meet tight schedules and years of experience in the field which can be further consolidated in a technology of interest to national industry. Close collaboration is still maintained of course with Garching.

On the other hand, the project W6, which has limited dimensions and requires the investigation of singular problems, is well sited at Garching.

APPLICATION OF SUPERCONDUCTING MAGNETS TO THERMONUCLEAR REACTORS. INFLUENCE OF MAGNET COST ON TOTAL PLANT COST

Some design studies of thermonuclear reactors operating with constant magnetic fields have been made (22). Since the magnetic fields involved are very high (100 kOe and more), conventional magnets are unsuitable owing to the enormous power losses, which can be avoided by using superconducting magnets (23).

By analogy with the total current density attained in large magnets such as those for bubble chambers of diameter 4 m and more, the total current density required for magnets of the dimensions necessary for thermonuclear reactors will probably not be larger than 25 A/mm^2. This is due to the stabilizing copper, to the necessary protection system, and, most of all, to the mechanical supporting structure in the magnet. In fact, the yield strength the superconducting composite is subject to in the winding is expressed by

$$\sigma_t = r \, J(r) \, H(r).$$

Therefore, assuming the product $J(r) \, H(r)$ to be constant, the volume of steel or similar material which reinforces the superconducting composite and which has the purpose of balancing σ_t must increase with increasing r.

In the case of magnets for thermonuclear reactors superconducting materials must have high H_c and relatively low J_c, the opposite being the case for magnets for plasma experiments. In this connection the importance of the compounds Nb_3Sn, V_3Ga, etc. and future materials having a still higher H_c is evident. The latter can be used more economically than V_3Ga and Nb_3Sn. It would also be desirable that such very high field materials be available in non-brittle form.

The superconducting materials in a thermonuclear reactor are subject to neutron irradiation. The relevant damage caused in superconductors have been the object of study for some time; however, the results have not yet shown a general connection between the metallurgy of the material and the damages derived due to neutron irradiation. Progress in this direction is desirable.

Of fundamental importance is also the influence of the magnet cost on the total cost of a reactor. A number of studies have been made which show the capability of a thermonuclear reactor of toroidal form to produce electricity economically [7]. Preliminary studies made at Culham on a 750 Mw reactor with mirror magnetic configuration show the competitivity of such a reactor in comparison with traditional systems, and that the superconducting magnet represents the major cost. Similar results are obtained in the study of a 2.500 MW toroidal reactor.[7]

As already observed, the cost of a superconducting magnet increases noticeably with increasing maximum field required owing to the considerable increase of the superconducting material volume which has to be used when the material has a maximum critical field close to the magnet maximum field. It is therefore evident that the cost of a reactor is governed, besides by the cost of the superconducting material, by the availability of superconductors having a critical field considerably higher than the maximum field required for the magnet in order to guarantee economical application.

CONCLUSIONS

The developments during the last years of high field superconducting materials and particularly adiabatic stabilization permit the construction of very reliable high field magnets.

Superconducting magnets are an extremely important tool for constructing plasma experiments and are a marked improvement on the traditional magnet for economy, smaller volume, and possibility of higher fields. This will become increasingly true since these experiments tend to have still larger dimensions and higher fields. The planning of the magnetic system for the stellarator W7 at Garching is a conspicuous example of this application.

The application of superconductivity offers moreover the possibility of constructing experiments not possible by traditional means, an example being the quadrupole W6 under construction at Garching.

The application of superconducting magnets is a key element in a possible future thermonuclear reactor. Since the magnet cost considerably affects the total cost of the plant, a reduction of the superconducting material cost is desirable, together with the development of very high field materials which would allow their economical use in producing the high fields required for thermonuclear reactors.

This work was performed under the terms of the agreement on association between the Max-Planck-Institut für Plasmaphysik and EURATOM.

REFERENCES

+ An earlier version of this paper was given in Italian at the "Tavola Rotonda sulle Applicazioni della Superconduttività", Frascati (Rome), February 16, 1971 LNF-71/25.

(1) - S. Foner et al., Phys. Letters $\underline{31A}$, 349 (1970)
(2) - K. Tachikawa and Y. Iwasa, Appl. Phys. Letters $\underline{16}$, 230 (1970)
(3) - R.V. Smith, Proc. 1968 Summer Study of Superconducting Devices and Accelerators (Brookhaven, 1968), Part I., p. 249
(4) - Z.J.J. Stekly, J. Appl. Phys. $\underline{42}$, 65 (1971)
(5) - M.N. Wilson et al., J. Physics $\underline{3D}$, 1517 (1970)
(6) - J.R. Purcell, Proc. 1968 Summer Study of Superconducting Devices and Accelerators (Brookhaven, 1968), Part I, p. 765
(7) - R. Hancox, Proc. 6th Symp. on Fusion Technology (Aachen, 1970), p. 83
(8) - W. Amenda et al., Fifth Symp. on Fusion Technology (Oxford, 1968)
(9) - R. Hancox, Proc. Third Intern. Conf. on Magnet Technology (Hamburg, 1970), in the press
(10) - G. v. Gierke and G. Grieger, IPP Festschrift 45 (1970)
(11) - W. Amenda and A.P. Martinelli, Paper to be presented at the XIII Intern. Congress of Refrigeration, Commission I, Washington, August 27 - September 3, 1971
(12) - F. Rau, Frühjahrstagung der Deutschen Physikalischen Gesellschaft, Fachausschuß Plasmaphysik, Bochum, 1971
(13) - F. Rau, Internal report IPP 2/67 (1968)
(14) - F. Rau, IPP Festschrift 49 (1970)
(15) - E. Berkl et al., Phys. Rev. Letters $\underline{17}$, 906 (1966)
(16) - J.D. Lawson, Proc. Phys. Soc. (London) $\underline{870}$, 6 (1957)
(17) - B. Oswald, Proc. Third Intern. Conf. on Magnet Technology, Hamburg (1970), in the press
(18) - L.W. Toth, Modern Plastics 123 (1965)
(19) - W.F. Westendorp and H. Hurwitz, General Electric report N. 63-RL (3254 E) (1963)
(20) - A.P. Martinelli, Proc. Third Intern. Conf. on Magnet Technology, Hamburg (1970), in the press
(21) - A.P. Martinelli, Proc. 6th Symp. on Fusion Technology, Aachen (1970), p. 101
(22) - Proc. BNES Nuclear Fusion Reactors Conf. (Culham Lab., 1969).
(23) - W. Kafka, Internal Report IPP 4/70 (1970)

APPLICATION OF SUPERCONDUCTORS TO MOTORS AND GENERATORS

Joseph L. Smith, Jr.

Massachusetts Institute of Technology

The development of high-field superconducting wire, suitable for fabrication into windings for electric machines, presents the opportunity for significant improvements in rotating electric machines. Previously copper and iron have been the obvious materials to use. In the evolution of the machines since the turn of the century, the most significant changes have been in the insulations and the coolants. Metallurgical improvements have also improved the performance of the magnetic circuits of the machines.

In contrast to the past evolution of machines, the advent of superconductors should motivate radical changes in the basic construction, especially in the largest machines. The potential of superconductors is best illustrated by examining the limitations of conventional machines and then evaluating how superconductors can advance these limits.

In all electric machines the electromechanical energy conversion results from an average magnetic shear stress between two electromagnetic components with an average relative mechanical velocity. Thus the machines are limited in power density by mechanical limits on the relative velocity and by electrical limits on the field strengths. In conventional machines the field strengths have been limited by the saturation of the iron which forms the magnetic circuit of the machine. High-field superconductors promise to extend this field limit by a factor of three and perhaps a factor of five. In addition, the high current density available in the superconductor allows a large magnetomotive force to be packed into a small volume without a significant energy cost.

In a conventional machine the current density in the conductor is limited by thermal degradation of the electrical insulation (temperature rise) or by the electrical losses (energy cost). The process of evolving conventional machines with higher magnetic shear stress in the air gap has required longer air gaps and correspondingly higher current densities in the windings. The higher current density has been provided by, first, hydrogen cooling (rather than air cooling) and second, by direct water cooling of hollow copper conductors. As a result of the effective cooling, conventional machines may be designed up to an economic optimization of power loss versus power density (capital cost). Preliminary design calculations indicate that the current density available with high-field superconductors may well provide an increase in the economic power density by a factor of five to ten.

The major limitation on the application of high-field superconductors in electric machines is the AC loss characteristics. Specifically, the significant AC loss at power frequencies limits the practical application to the windings which have essentially a constant flux linkage, i.e. the field windings. The configuration of a superconducting machine must be a superconducting field winding (rotating or stationary) and a normally conducting armature (AC) winding, which also may be stationary or rotating. If the armature is connected to the power terminals, the machine is a synchronous AC machine (Fig. 1). If the armature is connected through a commutator (mechanical rectifier/inverter), the machine is electrically a conventional DC machine (Fig. 2). A third alternate is the acyclic machine (homopolar or Faraday disk machine)(Fig. 3). In this machine a field winding produces the field and the electromechanical interaction occurs between an armature disk (half turn and infinite number

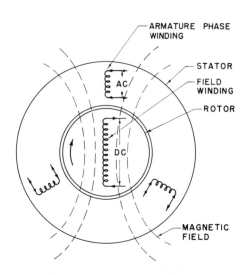

Fig. 1   Schematic for an AC Synchronous Machine

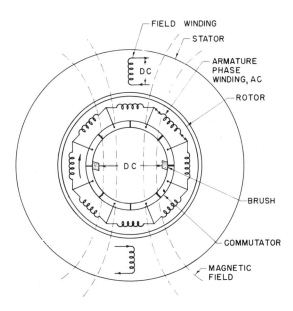

Fig. 2 Schematic for a DC Commutated Machine

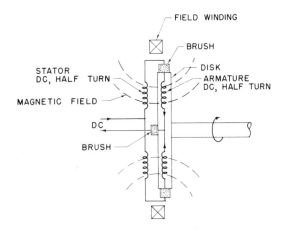

Fig. 3 Schematic for an Acyclic DC Machine

of phases) and a half-turn DC reaction winding (current leads to brushes).

The use of superconductors in the field windings of each of these machines provides an increase in the average field strength and a corresponding increase in the power density. However, the DC machines have limitations not related to the field strength. In each the full power of the machine must be transmitted across sliding contacts. The relative velocity at the brushes is limited by friction and/or wear to perhaps 200 ft/sec. In the commutated machine the voltage is limited by arcing to several hundred volts. In the homopolar machine the voltage is limited by the half turn windings, and the tip brushes must run at full air-gap relative velocity. However, multiple disk and segmented disk configurations and liquid metal brushes have advanced these limits to some extent.

The application of superconductors in the field winding of a commutated DC machine has been studied at the Leningrad Polytechnical Institute and an experimental machine has been constructed (1).

The application of superconductors in the field winding of a homopolar machine is especially practical since the field winding is not directly involved in the electromechanical interaction of the machine and thus is not required to sustain any mechanical reaction forces, but it still must withstand its own self forces. In addition the machine can employ the well developed technology of construction of superconducting solenoids. The homopolar machine has been studied thoroughly at International Research & Development Company in England. A preliminary 50 hp machine was constructed (2) and more recently a 3000 hp motor has been successfully tested (3,4). The British are now studying the potential of the superconducting homopolar machine for ship propulsion.

In contrast to the DC machines, the AC synchronous machine has the armature directly connected to the terminals and therefore can utilize high voltages, and modest currents achieve ratings as large as 1200 MW with conventional iron and copper construction. These large machines are mechanically limited by rotor tip velocity (centrifugal stress) and by rotor lateral flexability (critical speeds). In the absence of electrical limitations the use of superconductors in the field winding offer the possibility of significant increases in machine rating within the mechanical limitations. The application in large machines is most promising because the cryogenic refrigeration equipment becomes a smaller part of the superconducting machine cost (5) as the size of the machine is increased.

Work on superconducting alternators was started by Woodson, Stekly, et.al. (6,7) with the construction of a demonstration alternator with a stationary field and a rotating armature, connected through slip rings. Early experimental work by Oberhauser and Kinner (8) showed that superconducting armatures and air gap shear in a fluid at or near $4.2°K$ are not desirable features of a superconducting machine. The dissipation associated with shearing an air gap fluid (even a gas at low pressure) involves serious problems in maintaining superconducting temperatures and requires excessive power for refrigeration.

APPLICATION TO MOTORS AND GENERATORS

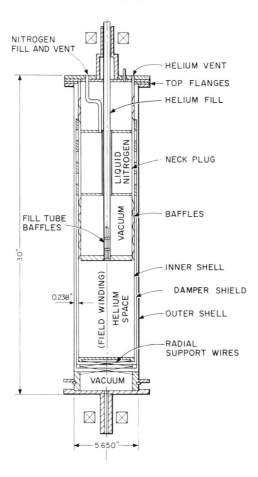

Fig. 4 Rotor with Superconducting Field Winding

The work on alternators at MIT had as its initial objective the demonstration of the practicability of a rotating field winding with no air-gap shear at helium temperature. The combined efforts of the Cryogenic Engineering Laboratory and the Electric Power Systems Laboratory at MIT were brought to bear on the problem with the support of the Edison Electric Institute. The rotor shown in Fig. 4 has very successfully produced a dipole field with a superconducting winding rotating at 3600 RPM. The superconductors, the liquid helium coolant and the complete dewar vessel all rotate as a unit. The dewar is basically a conventional wide mouth dewar with precision alignment for balance and bicycle spokes at the bottom for lateral support. The neck closure of the dewar, which carries the winding, must be vacuum insulated since the normal thermal stratification of the gas in the neck fails in the centrifugal field. The details of the rotor are given by Thullen (9,10).

Experiments with the rotor of Fig. 4 showed no basic obstacles to the realization of a rotating superconducting field winding. Specifically the spin-up of the liquid helium and mechanical vibrations during rotation do not cause excessive evaporation of liquid helium. As expected, rotation does not adversely affect the characteristics of the superconductor. The experiment did however point out the extraordinary effectiveness of natural convection heat transfer at cryogenic temperatures in a 1000 g centrifugal force field.

An armature, Fig. 5, was designed to take advantage of the high field produced by the rotor. At first it may seem that applying superconductors to a synchronous machine involves only the field winding since the armature cannot be superconducting. However, a conventionally designed armature cannot take advantage of the full potential offered by the superconducting field. Basically, the steel is not needed in the magnetic circuit, and if it is present, it will be highly saturated resulting in high losses. Thus the laminated steel teeth normally found in a machine are not included in the armature design. This immediately provides about twice as much space in the active region for working armature conductors; however, it also eliminates the slots which normally support the conductors. In addition, the armature conductors are now exposed to the full AC field which normally is in the iron teeth. This places the mechanical reaction forces which are normally on the iron teeth, directly

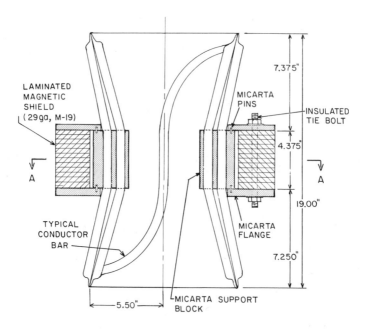

Fig. 5 Armature for Experimental Synchronous Machine

APPLICATION TO MOTORS AND GENERATORS   489

on the conductor. Also in order to avoid eddy-current losses (due
to skin effect) and circulating current losses (due to the inhomogeneous magnetic field), the armature conductor must be composed of
fine strands which are transposed. The armature of Fig. 5 (11)
employs round Litz wire for the conductors which are wedged between
plastic teeth.

Although the machine of Fig. 6 (12) is an air core machine, it does
have a laminated steel ring around the armature. This ring-shield
confines the rotating dipole field within the machine to avoid para-
sitic losses in conducting material in the vicinity. The shield also
provides a symmetric boundary condition and avoids pulsating forces
on the rotor.

The machine was constructed with a vertical axis so that it
could be filled with liquid helium and then run until the liquid
boiled away. After a series of successful batch-fill runs, a vacuum
insulated transfer tube, Fig. 6, was constructed with a rotatable
joint. With this apparatus the machine has been run for several
hours and could be run continuously if coupled directly to a helium
liquefier.

Fig. 6   Experimental Machine

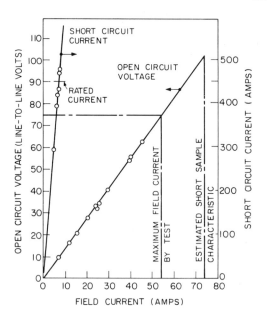

Fig. 7 Experimental Machine Characteristics

The steady-state electrical characteristics (13) of the machine were measured in open-circuit and short-circuit tests. The basic results are shown in Fig. 7. These results confirm the high power density possible in superconducting machines, in spite of the non-optimum design of the experiment. The field winding achieved 3.2 Tesla both rotating and stationary. This is about 75 percent of the short sample capability of the wire. More recently the machine has been run as a synchronous condenser connected to the laboratory power supply. The machine has achieved 45 KVA overexcited. A typical V-curve is shown in Figure 8.

After success with the initial experiments with the first machine, the MIT group began the analysis of transient and fault behavior of superconducting turbine generators connected in an electric power system. The electrical characteristics of full-scale machines of preliminary design were developed (5). Einstein (14) then used these characteristics to investigate the interactions between the superconducting machine and the power system during a fault on a transmission line and the subsequent action of the circuit breaker to clear the fault. He concluded that the superconducting machine can operate in a stable manner while connected to a power system which experiences normal disturbances.

Luck (15) studied the internal behavior of superconducting alternators during severe fault conditions. He analyzed in detail

APPLICATION TO MOTORS AND GENERATORS

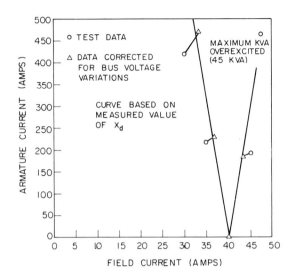

Fig. 8 Experimental Machine V Curve

the currents, magnetic fields, and mechanical forces which result from faults. He concluded that the preliminary designs did not have sufficient mechanical strength to withstand the fault torques which are as high as ten times the full load torque. He devised a rotating, room temperature, cylindrical shield for the rotor which overcomes this problem without introducing an unacceptable heat leak into the cryogenic parts, Fig. 9.

The group at MIT is now concentrating on the construction of a second experimental machine, Fig. 10. The objectives of this experiment are: to demonstrate a design and methods of construction which take maximum advantage of the characteristics of the superconducting field, to verify the analysis of the machine characteristics, to verify practical design equations, and to provide experience and background for a decision to build a full scale prototype machine.

The size of the machine was selected as the largest experiment within the capabilities of the Laboratory facilities. The preliminary estimate of the machine rating is 2 MVA. The superconducting rotor is eight inches in diameter with a 24-inch active length. With the thermal isolation tubes, seals and bearings the length is about 6 feet. The six layers of superconducting wire are bound down to the central tube with fiberglass and epoxy bands. The liquid helium enters through the hollow shaft, cools the winding, cools the midpoint of the thermal isolation tubes and leaves through the same

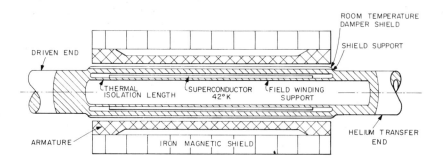

Fig. 9   Superconducting Rotor for 1000 MVA Machine

shaft. A double rotatable transfer coupling connects the rotor to the liquid supply dewar.

The rotor turns in a plastic high-vacuum chamber and the shafts are sealed with oil buffered face seals. The rotor is surrounded by a cold copper shell which is supported and cooled at the mid point of the thermal isolation tubes. This shell serves as a thermal radiation shield and as an electrical shield to prevent AC fields from reaching the superconductors.

The three-phase two-pole armature is composed of 228 conductor bars packed solidly around the central plastic tube (also the vacuum chamber) and bound down with three layers of continuous-filament fiberglass roving and epoxy resin. Each bar is composed of eighteen seven-strand cables Roebel transposed in a rectangular cross section. The epoxy insulation around the bar is cured after the bars are in place in the machine so that the winding is a monolithic structure of copper, fiberglass and epoxy. The bars are in two layers in the active length and in four layers in the ends to provide space for the turns to the end connections between the bars. The winding is connected so that adjacent bars are in electrical sequence. With this connection the bar insulation is only subjected to turn voltage. The mechanical forces on the armature winding are carried to the end flanges by a fiberglass reinforced torque tube wound over the outside of the armature. The windings are cooled by transformer oil which is pumped through and around the armature. The oil fills the outer case of the machine and thus also cools the flux shields.

The armature winding fits into the bore of the laminated magnetic-steel flux shield. The shield extends out over the end turns of the windings. The laminated ferro-magnetic shield in turn fits

APPLICATION TO MOTORS AND GENERATORS

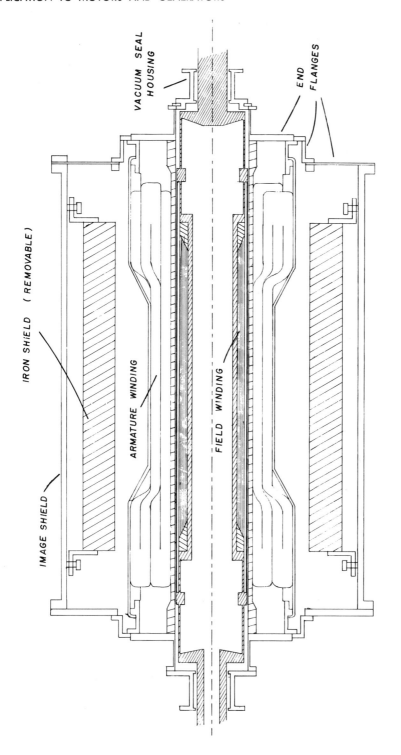

Fig. 10 Cross Section of MIT - EEI Experimental Machine

inside the aluminum outer case for the machine. The machine is constructed so that it can be assembled and operated with the magnetic shield left out. The eddy currents in the outer case then form an image flux shield for the machines. The use of the image shield reduces the cost and weight of the machine but increases the losses. Data from tests on the machine will provide the basis for selection of the method of shielding; however, preliminary calculations indicate an economic advantage for the image shield in large machines.

In conclusion, the studies by the MIT superconducting machine group have shown that the greatest potential for the application of superconductors is in the field windings of large synchronous generators. Machines which are optimized to take full advantage of the potential promise the following advantages:

(1) Higher power density will provide smaller, lower cost machines and make possible machines of larger capacity.

(2) Lower machine reactances will improve machine stability in the power system.

(3) Heavy and expensive laminated steel is eliminated from the magnetic circuit.

(4) Higher economic terminal voltages for the machine are possible.

(5) The superconductor reduces exciter power requirements and increases the efficiency.

References

1. Kazovskiy, Kartsev and Shaktarin, <u>Superconducting Magnet Systems</u>, Science Publishing Company, Leningrad, 1967, pp. 186-190 (in Russian).

2. A. D. Appleton and R. B. MacNab, "A Model Superconducting Motor" Commission I, London, "Annex 1969-I Bulletin International Institute of Refrigeration", pp. 261-267.

3. A. D. Appleton and J. S. H. Ross, "Aspects of a Superconducting Winding for a 3 250 hp Motor". Commission I, London, "Annex 1969-I Bulletin International Institute of Refrigeration", pp. 277-283.

4. F. Tinlin and J. S. H. Ross, "The Cryostat and Refrigerator for a 3 250 hp Superconducting Motor". Commission I, London, "Annex 1969-I Bulletin International Institute of Refrigeration, pp. 277-283.

5. H. H. Woodson, J. L. Smith, Jr., P. Thullen, and J. L. Kirtley, Jr., "The Application of Superconductors in the Field Windings of Large Synchronous Machines", "IEEE Trans. Power Apparatus and Systems," Vol. PAS-90, No. 2, March/April 1971, pp. 620-627.

6. H. H. Woodson, Z. J. J. Stekly, and E. Halas, "A Study of Alternators with Superconducting Field Windings: I-Analysis," "IEEE Transactions on Power Apparatus and Systems," Vol. PAS-85, No. 3 March, 1966, pp. 264-274.

7. Z. J. J. Stekly, H. H. Woodson, A. M. Hatch, L. O. Hoppie, and E. Halas, "A Study of Alternators with Superconducting Field Windings: II-Experiment," "IEEE Transactions on Power Apparatus and Systems," Vol. PAS-85, No. 3, March, 1966, pp. 274-280.

8. C. J. Oberhauser and H. R. Kinner, "Some Considerations in the Design of a Superconducting Alternator," "Advances in Cryogenic Engineering," Vol. 13, K. D. Timmerhaus, Editor, 1968.

9. P. Thullen and J. L. Smith, Jr., "The Design of a Rotating Dewar and Field Winding for a Superconducting Alternator," Advances in Cryogenic Engineering," Vol. 15, K. D. Timmerhaus, Editor, 1970, pp. 132-140.

10. P. Thullen, "Analysis of the Application of Superconductivity to Commercial Electric Power Generation," Sc.D. Thesis, Department of Mechanical Engineering, MIT, Cambridge, Massachusetts, June, 1969.

11. J. C. Dudley, "Fabrication of an Armature for a Generator with a Superconducting Rotating Field Winding," SM Thesis, Department of Mechanical Engineering, MIT, Cambridge, Massachusetts, May, 1969.

12. P. Thullen, J. L. Smith, Jr., and W. D. Lee, "The Design and Testing of a Rotatable Liquid Helium Transfer Coupling for Use at 3600 RPM,". Presented at XIII Congress of Refrigeration, International Institute of Refrigeration, Washington, D.C., August 27-September 2, 1971

13. P. Thullen, J. C. Dudley, D. L. Greene, J. L. Smith, Jr., and H. H. Woodson. "An Experimental Alternator with a Superconducting Rotating Field Winding", "IEEE Trans. Power Apparatus and Systems," Vol. PAS-90, No. 2, March/April 1971, pp. 611-619.

14. Thomas H. Einstein, "Excitation Requirements and Dynamic Performance of a Superconducting Alternator for Electric Utility Power Generation", Sc.D. Thesis, Department of Mechanical Engineering, MIT, Cambridge, Massachusetts, December, 1970.

15. David Lee Luck, "Electromechanical and Thermal Effects of Faults Upon Superconducting Generators", Ph.D. Thesis, Department of Electrical Engineering, MIT, Cambridge, Massachusetts, June, 1971.

SUPERCONDUCTING COILS

Z. J. J. Stekly

Magnetic Corporation of America

Cambridge, Massachusetts 02142

INTRODUCTION

The ability of Type II superconductors to maintain zero electrical resistance under direct current conditions while in the presence of high magnetic fields has led to various current and proposed future uses for superconducting coils. These include high energy physics applications, containment coils for controlled thermonuclear fusion research, saddle shaped coils for energy conversion in magnetohydrodynamic power generation, superconducting coils for the field windings of electrical motors and generators, both for utility applications as well as for electric ship propulsion, levitating coils to suspend a high speed ground vehicle above an electrically conducting guideway, coils for magnetic storage of electrical energy, as well as laboratory coils for use in a wide variety of research ranging from low temperature solid state physics to novel applications of the magnetic fields.

In all of the above applications, the heart of the system is a superconducting coil which generates a magnetic field. While in many instances these coils bear little physical resemblance to one another, nevertheless they have much in common. It is these common characteristics of superconducting coils that will be the main topics discussed below.

The material presented is the basis for current superconducting coil technology. Consideration is given first to the characteristics of the superconducting materials themselves and the reasons for the particular configuration of the composite conductors used to wind the coils. Following this, the initiation and propagation of a normally conducting region and means for protection of the coils are treated. These considerations affect the design current density of a particular coil. Values of current density for state of the art coils are reviewed in terms of coil design, stability of the conductor, and quench problems. This is followed by a discussion of the structural requirements to support the magnetic loads.

The current state of the art of superconducting coils is then discussed in terms of size, magnetic field, and energy storage capabilities.

SUPERCONDUCTORS

The basic ingredient of superconducting magnets is the superconductor itself. Although the number of superconducting materials is very large, up to now only Nb-Zr, Nb-Ti, and the compound Nb_3Sn have been used in magnets to any significant extent.

Nb-Zr and Nb-Ti are alloys, are ductile, and can be made into conductors by standard metal working techniques. Historically, Nb-Zr was widely used until about 1967; since then, the more ductile Nb-Ti alloys have replaced Nb-Zr.

The critical temperatures of the conductors of the NbTiZr system are in the neighborhood of 10K. These materials are restricted to fields below 100 kilogauss at 4.2K.

Nb_3Sn is a brittle material; however, it has a critical temperature of 18K, and remains superconducting in fields up to 245 kilogauss at 4.2K. Due to its brittle nature, it has been made into conductors by vapor deposition on a steel substrate or by reacting a thin layer of tin with a niobium substrate.

Recently V_3Ga, another brittle compound, has become commercially available in conductor form. This material is comparable to Nb_3Sn and has a critical temperature of 16.8K and a critical field of 210 kilogauss at 4.2K.

Figure 1 shows typical curves of current density versus magnetic field for the superconductor itself at 4.2K for some of the ductile Nb-Ti alloys and Nb_3Sn.

Other materials with potential future use in superconducting coils are $Nb_3(Al_{0.8}Ge_{0.2})$ and Nb_3Al, all brittle compounds. Table I summarizes the useful (or potentially useful) magnetic field range and the critical temperature for these materials.[1]

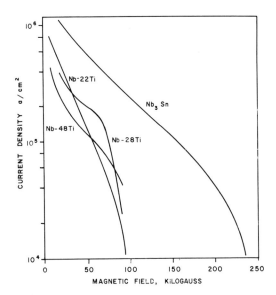

Figure 1. Curves of superconductor current density for three ductile alloys of Niobium and Titanium as well as for the brittle compound of Niobium and Tin, Nb_3Sn.

Table I. Superconductor Properties

Material	Critical Temp. (K)	Critical Field At 4.2K (kG)
Brittle Compounds		
Nb_3Sn	18.2	245
V_3Ga	16.8	210
$Nb_3(Al_{0.8}Ge_{0.2})$	20.7	410
Nb_3Al	17.5	300
Alloys		
Nb-48% Ti	9.5	122

STABILIZATION OF THE SUPERCONDUCTOR

In addition to being mechanically sound, conductors must operate reliably in the magnet windings. In order to provide stability against disturbances of electrical, thermal or mechanical nature which may occur in the windings, a material of high electrical and thermal conductivity is included in the conductor. Copper is used primarily in the alloy system. Silver and copper have been used to stabilize Nb_3Sn conductors. Although it has a higher electrical conductivity than copper (especially at high magnetic fields), high purity aluminum has found only limited use as an integral part of the conductor because of the problem of making a metallurgical bond with the superconductor.

The question of stability of operation of a superconductor can best be understood by referring to Table II,[2] which shows some thermal and magnetic properties at 4.2K for three superconductors most widely used. For comparison, properties for copper and for liquid and gaseous helium are included in the Table wherever applicable.

Table II. Thermal and Magnetic Properties

			Superconductors			Copper	Helium	
			Nb_3Sn	Nb-Zr	Nb-Ti		Liq.	Gas
Thermal Conductivity	k	(mw/cmK)	0.4	0.8	1.2	7000.	2.72	0.1
Specific Heat	c	(mJ/gK)	0.21	0.18	0.18	0.099	4480.	6000.
Mass Specific Heat	γc	(mJ/cm^{3}K)	1.13	1.46	1.01	0.89	560.	48.
Normal Resistivity	ρ_n	($\mu\Omega$-cm)	26.	34.	24.	0.03	...	...
Density	γ	(g/cm^3)	5.4	8.1	5.6	8.9	0.125	0.008
Critical Temperature	T_c	(K)	18.	10.5	9.5	...	...	...
Thermal Diffusivity	$k/\gamma c$	(cm^2/sec)	0.35	5.4	1.18	7900.	0.005	0.002
Electrical Diffusivity	ρ_n/μ_o	(cm^2/sec)	2069.	2706.	1910.	2.5	...	...
Ratio of Diffusivities	$k\mu_o/\gamma c \rho_f$	($H/H_{c2}=0.1$)	0.00169	0.020	0.0062	7600.	...	...

SUPERCONDUCTING COILS

From Table II the following conclusions can be drawn:

1. Thermal conductivity of superconductors is very poor compared with copper.

2. Of the materials listed, the only significant specific heats (volumetric or by weight) are those of the helium.

3. The thermal diffusivity of copper is much higher than that of the superconductors.

4. The magnetic diffusivity is much higher for the superconductors in the flux flow regime than for copper.

5. The ratio of thermal diffusivity to magnetic diffusivity is much less than unity for superconductors and much larger than unity for copper.

As a consequence of Item 5 above, magnetic field changes in a superconductor take place much more rapidly than heat can be conducted away by thermal diffusion. Conversely, for copper (at 4.2K) thermal diffusion takes place much faster than magnetic field changes. Another way to state the effect is to say that for superconductors the magnetic field changes take place adiabatically with no heat flow during the change. Copper on the other hand goes through a series of thermal steady states as the magnetic field varies.

It is this difference in the behavior of the materials which makes copper (or aluminum) an effective stabilizing substrate.

The fact that liquid helium has considerable specific heat makes it a very useful heat sink in the design of magnets.

The simplest and most conservative stability criterion is based on a steady state analysis of the conductor terminal characteristics as determined by the superconductor, the stabilizing substrate, and the heat transfer from the conductor to the liquid helium. The terminal characteristics are arrived at by equating the heat generated by the conductor to the heat transferred from the conductor to liquid helium.[3,4,5,6] A simplified analysis of the steady state stability is presented in Appendix A. The main results of the approach are that:

1. The temperature in the conductor with all the current flowing in the substrate must be less than the critical temperature (at zero current and at the operating magnetic field) of the superconductor itself. Due to the non-linear character of heat transfer to liquid helium, this requires that conductors

be designed for a maximum heat flux per unit area of conductor exposed to liquid helium with all the current in the substrate. For small passages (of the order of 0.2 - 0.4 mm) the accepted value is approximately 0.3 w/cm². Larger cooling passages allow higher values to be used.

2. The diameter (round strands) or thickness of the superconductor should be less than a certain size in order that the temperature drop in the superconductor be small. This size is given in Appendix A as:

$$d < 2r_{cs} = \frac{2}{j_{ch}} \left[\frac{k \, (T_{ch} - T_b)}{\rho} \frac{A}{A_s} \right]^{1/2} \tag{1}$$

where:

k - thermal conductivity of the superconductor

ρ - resistivity of the substrate material

T_{ch} - critical temperature at zero current in the presence of a magnetic field

T_b - bath temperature

j_{ch} - superconductor critical current density in the presence of a magnetic field

$\frac{A}{A_s}$ - ratio of substrate area to superconductor area

Due to the relative simplicity of the model analyzed, the above limit of superconductor size should be interpreted as being only an approximate value. The strand size is field dependent, temperature dependent, and geometry dependent because of its variation with superconductor current density, critical temperature, the ratio of substrate area to superconductor area, and the resistivity of the substrate.

Typical computed values for Nb-Ti and copper composites yield values for strand sizes of the order of 50 to 100 microns.

The cooling requirement and the need for substrates in coils designed for steady state stability generally result in overall current densities in the winding of less than 10,000 A/cm².

Although coils designed for steady state stability are reliable and the coil performance can be accurately predicted, the need for higher current densities exists for reasons of compactness, weight or cost. Internal cooling passages in a winding lead to more complex and consequently more costly construction; consequently, their partial or total elimination is very desirable.

The initial approach to increasing current density was empirical. Current density coils were designed primarily on past performance. Although the empirical nature of the design has not been entirely superseded, the stability criteria for design have been established both analytically and experimentally and most currently available conductors are capable of operating at current densities above these based on steady state stability.

The stability criteria for these conductors are based on dynamic analyses of the flux motion through a conductor.

The first of these criteria [7,8] is called dynamic stability, and is based on surrounding the superconductor with enough copper or other highly conducting (both electrical and thermal) material to slow down any penetration of flux into the superconductor to such an extent that the heat generated whithin the superconductor by the motion of flux can flow into the substrate and be conducted away.

When the criterion is satisfied, the flux flow into the superconductor is slowed down so that the superconductor reaches a thermal steady state. The resulting design criteria result in the same condition for strand size as for steady state stability [Equation (1)]; however, there is no longer the need for the conductor to be cooled along its entire length.

Most recently attention has been concentrated on the superconductor itself, and heat transfer to the substrate is neglected. The result of this analysis is a stability criterion referred to as intrinsic or adiabatic stability.[9] The stability criterion is determined by postulating a small change in flux, computing the temperature rise, and determining whether the resulting current distribution (associated with a flux change) is compatible with the temperature rise.

The result of this criterion is:

$$d < \frac{\pi}{2j_{ch}} \left[\frac{\gamma c (T_{ch} - T_b)}{\mu_o} \right]^{\frac{1}{2}} \quad (2)$$

where:

$$\mu_o = 4\pi \times 10^{-7} \text{ (H/m)}$$

γ = density (kg/m^3) of the superconductor

c = specific heat (joules/kgK) of the superconductor

An analysis[10,11] for finite size flux penetrations (rather than one based on infinitesimal ones) results in larger allowable superconductor size.

Based on the above criteria, it is evident that the following lead to stability.

1. Small superconductor size.

2. Addition of copper or another highly conducting substrate.

3. Cooling with liquid helium.

The most desirable conductor is then one which has the superconductor divided up into many small strands. These points were all observed experimentally before the analytical basis became evident.[12]

There is one more important element in the stability picture – the twisting of the conductor.[13]

This effect is best understood by referring to Figure 2, which shows the cross section of a conductor containing only two superconducting strands in an electrically conducting substrate. The conductor is exposed to a uniform time varying magnetic field (transport currents are neglected). The picture shows the field distribution for the actual geometry as well as for a developed geometry which shows the magnetic field perpendicular to the plane of the two strands.

The induced current is a shielding current and flows in opposite directions in each of the two strands. Only the component of magnetic field perpendicular to the plane of the strands induces circulating currents. The magnitude of the circulating current density in each superconducting strand is:

$$j_s = \frac{\ell^2 \frac{dB}{dt}}{\pi^2 \rho d_s} \qquad (3)$$

SUPERCONDUCTING COILS

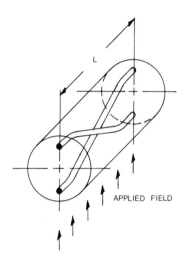

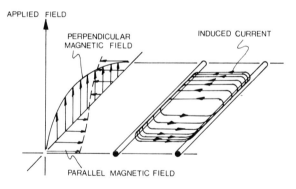

Figure 2. Effect of twist on the superconductor.

where:

j_s = induced current density in each superconducting strand

ℓ = length for 180° of twist

ρ = resistivity of the substrate

d_s = diameter of the superconducting strand

B = magnetic field

If the induced current density is comparable to the critical current density then the effective current carrying capacity is reduced. The induced eddy currents decay after the magnetic field stops varying in time. The decay time of these currents is given by:[9]

$$\tau = \frac{\mu_o \ell^2}{\pi^2 \rho} \quad (4)$$

It can be concluded that untwisted conductors ($\ell \to \infty$) have shielding currents that do not decay in time. On the other hand, even twisted conductors behave as though they were not twisted for high rates of change of magnetic field. An effectively twisted conductor allows the current to distribute evenly among all the individual strands in the composite and contributes very significantly to the overall stability.

Now that the basic conductor characteristics have been discussed, we can review several typical conductors.

AVAILABLE CONDUCTORS

Typical of the technology of NbTi conductors are the cross sections shown in Figures 3, 4, and 5.

Figure 3 shows low current density conductors with finely divided superconducting strands. These conductors have been designed based on steady state stability criteria[14] and are intended for use in large coils, 4.6 m diameter, in which the windings operate at 51 kilogauss. The use of the steady state criteria has been common in large coil systems where reliability and predictability outweigh current density considerations.

SUPERCONDUCTING COILS

Figure 3. Proposed conductors for use in a large Bubble Chamber at Centre Européen de Recherches Nucléaires (CERN). The conductors are required to carry 6450 A at a current density of 3500A/cm^2 at 50 kilogauss. Conductors are copper or aluminum with Nb-Ti strands. (Courtesy of CERN.)

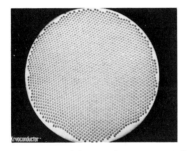

Figure 4. Cross section of 1.82 mm diameter Nb-Ti composite containing 2133 strands of Nb-Ti in a copper matrix with 0.5 twists/in. Capable of carrying 2200 A at 60 kilogauss (j = 5.2 x 10^4 A/cm^2).

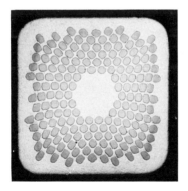

Figure 5. Cross section of Nb-Ti-Copper Composite, 1.62 mm square, 180 strands of Nb-Ti, 1 twist/in. The inner strands have not been put in because they do not twist. This conductor is capable of carrying a current of 1050 A at 60 kilogauss (j = 4 x 10^4 A/cm^2).

Figure 4 shows the cross section of a conductor at the other end of the current density scale. The particular picture shows a conductor 1.82 mm in diameter with 2133 finely divided superconducting strands. Conductors of this type can be made in a large range of sizes and shapes.

The general ease of fabrication of Nb-Ti and copper composites makes possible a large variety of shapes and sizes. Round, square or rectangular, and hollow conductors have been made. Conductor sizes have ranged from 50 microns to a fraction of a cm in size. The superconducting strand size in general varies with overall conductor size - strand sizes of 50 to 100 microns are typical of conductors between .75 mm to 2.5 mm. Larger conductors have larger strand sizes. The number of twists also varies with conductor size and can be up to one twist per cm for small diameter conductors.

Figure 5 shows a conductor of square cross section where the central strands have been left out because they cannot be twisted.

Although Nb_3Sn has a higher current density and higher critical magnetic field than Nb-Ti, its use has been limited, because of its brittle nature, to magnetic fields of 100 kilogauss and above where the Nb-Ti conductors are useless.

Due to its brittle nature, Nb_3Sn cannot be made into conductors by standard metal working processes. Up to now two methods of manufacture have been carried through to the commercial stage: (A) The plating of niobium ribbon with tin and then reacting to form a thin layer of Nb_3Sn (Copper ribbon is generally soldered on for stability.); (B) The vapour deposition of Nb_3Sn onto a stainless steel substrate (usually provided with silver plating or copper plating for stabilization.)

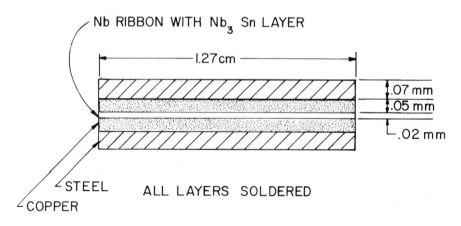

Figure 6. Reinforced Nb_3Sn conductor.

Figure 6 shows the cross section of a Nb_3Sn conductor used in the high field windings of an 8.8T, 50 cm bore magnet.[15,16] This particular conductor required the addition of reinforcing stainless steel for structural purposes.

CONDUCTOR AND COIL PERFORMANCE

One of the standard tests performed on a conductor is to take a length of the order of a meter, place it in an external magnetic field which can be varied independently, and measure its characteristics as a function of magnetic field. This is done by placing voltage taps on the sample and measuring the sample voltage versus sample current at each value of external magnetic field.

Figure 7 is the result of several such tests, each performed at a different value of magnetic field, on a square conductor 2mm on a side with six 0.38 mm Nb-Ti strands in a copper matrix.[17] This size strand is large enough to exhibit strand size effects at the lower magnetic fields. The conductor was exposed to liquid Helium along its length.

The following behavior is observed as the current is increased. The voltage across the sample remains zero until the critical current of the superconductor is reached. At this point, a jump occurs, due to large strand size, but does not occur at higher magnetic fields. If the current is reduced after the jump, a recovery to fully superconducting behavior occurs. Further increasing the current past the initial jump (if any) results in a gradual increase in voltage with increasing current. In this regime, the current flows partly in the superconductor and partly in the substrate.

As the current is further increased, the conductor heats up and the superconductor carries less and less current. At the point labeled breakaway, the limit of nucleate boiling is reached. At this point, transition from nucleate to film boiling takes place accompanied by a large temperature rise and the current is all transferred to the copper substrate.

If the current is now lowered, the current will remain in the normal substrate until the recovery point is reached. The recovery point corresponds to the transition of film to nucleate boiling. At low magnetic fields, recovery currents are below the critical value. At high magnetic fields, the recovery current is higher than the critical current.

Below the recovery current there is only one stable operating condition - at zero voltage with all the current in the superconductor. Between the recovery point and the critical current there are other operating points in addition to the zero voltage one.

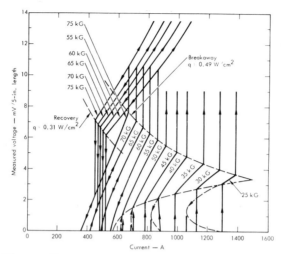

Figure 7. Terminal characteristics for a 2mm square conductor with six 0.38 mm strands of Nb-Ti.

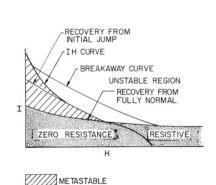

Figure 8. Conductor and coil winding stability map.

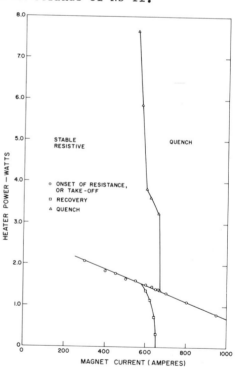

Figure 9. Response of a superconducting coil to local heat source within the windings.

SUPERCONDUCTING COILS

Where the conductor operates depends on the disturbances it is subjected to and its inherent stability. Rapid charging, mechanical vibration, shorts in the windings could result in expulsion of current from the superconductor into the substrate.

Data similar to that of Figure 7 are generally taken for samples that are mounted to simulate a particular coil construction. Data of this type can be summarized on a stability map which has the general characteristics shown in Figure 8.

Shown in the plot are several curves. The I-H curve represents the limit of the current carrying capacity of the conductor as determined by the amount of superconductor. The other curves represent demarkation lines for various types of behavior.

At low currents, steady state stable operation at zero resistance occurs under almost any disturbance provided helium permeates the winding. The limit of this region is determined by the transition from film to nucleate boiling at the conductor helium surface. Operation above the limit of the stable zero resistance region is metastable in the sense that a large enough disturbance will trigger a self-propagating resistive region. Since most disturbances are of a transient nature, the stability of the conductor to transients is the key to successful operation in this region. (The regions lying outside the I-H curve are all resistive and are of little practical interest.) The curve labeled "recovery from initial jump" is the strand size effect discussed above.

Once a coil has been built, its behavior or stability can be measured by means of a heater within the windings. Figure 9 demonstrates the behavior of a coil wound with multi-strand Nb-Ti conductor.[18] The heat generated simulates a rapid flux change, a poor joint, or any other disturbances resulting in local heat generation within the winding.

At low magnet current, no resistance appears across the coil terminals as the heater power increases until the line of onset of resistance is reached. If the heater power is increased further, a steady, non-propagating resistive region is created around the heater, which increases and decreases with heater power.

At currents above about 650 A, the behavior is quite different. No resistance appears until the onset of resistance is reached; however, once resistance appears, it is self-propagating, and a quench occurs. Reducing the heater power does not have any effect once resistance has appeared.

The region below 650 A is statically stable and the region above 650 A is metastable. The region below 650 A corresponds to the steady state stability design criteria already discussed. For the particular coil in question, the maximum current at which it could operated exceeded 1000 A, and was equal to the full critical current of the material used.

Normally a superconducting magnet should not be operated in a manner where it is likely to develop a self-propagating resistive region (quench). However, unforeseen conditions may occur that could result in a quench, and magnets must be designed to withstand this without damage.

The severity of a quench depends very strongly on the conductor current density. Damage during a quench may occur from one of two causes: too long a decay time, so that overheating of the initial normal spot occurs, or too high a voltage during the decay, so that arcing occurs in the windings.

A measure of the severity of the heating of the original normal spot is shown in Figure 10, which shows the temperature at the end of the quench as a function of initial current density and the time constant of the current decay which is assumed to be exponential. (If the current were to remain constant, the time to achieve the same temperature is one-half of that shown in Figure 10.)

It is evident that for high current density coils the required decay times can be quite short.

The voltage as well as the temperature in a quench increases with increasing coil energy,[19] so that there is a limit to the energy (which depends on the current density) which can be safely dissipated in an unprotected coil during a quench.

Larger coils generally tend toward lower current densities because of the requirement for increased stability and to minimize the problems of a quench.

Protection of the coil during a quench can be achieved by several means:

1. Use of an external resistor in which most of the coil energy is dumped. This system requires detection of a propagating normal region and performing the necessary switching. It has the advantage of dissipating a large fraction of the coil energy external to the cryogenic container, thus alleviating the problem of rapid helium boiloff.

2. Use of shunts to break up the coil into several smaller units so that the energy is distributed more evenly in the windings.

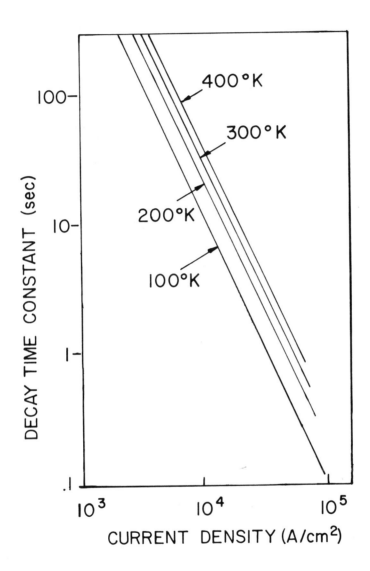

Figure 10. Temperature at the end of a quench as a function of initial current density and quench time.

3. Use of a circuit of normal conductor magnetically coupled to the superconductor. The time constant of the protective circuit should be much longer than the decay time of the superconducting current in order that effective protection results. This method has the advantage that the current is quickly removed from the superconducting circuit.

Considerations of stability and quench problems generally determine the overall current density in the windings for a particular application. In order to illustrate the general state of the art, the overall current density has been plotted in Figure 11 for coils larger than 7.6 cm diameter on which detailed data was available.[20] The data points are plotted as overall current density versus the maximum field at the windings.

The shaded region is an approximate demarcation line between areas where coils can be designed for static stability (below the shaded line) and where dynamic or intrinsic stability must be relied upon (above the shaded line).

The three coils having the highest current densities in the region below 60 kilogauss have an internal diameter of 9 cm.[9] They use twisted multiple strand copper and NbTi of .035 cm diameter with 60 filaments approximately .003 cm diameter with a copper to superconductor ratio of 1.1 to 1. The coil is impregnated with paraffin wax and has no helium within the windings.

Most of the coils below the shaded region are stable based on steady state stability criteria. Above the shaded region, although in many instances helium percolation exists in the windings, the coils operate in the metastable region.

The Nb_3Sn coils rely mostly on copper to damp any disturbances, while the NbTi coils rely on small strands, twisted conductor, and in some instances on helium percolation.

With few exceptions, all high current density coils above the shaded region are relatively small, in terms of stored magnetic energy. A sound technical basis has been established for high current density magnets; nevertheless the demonstration of high current densities in large sizes still remains to be done.

STRUCTURE

One of the most important aspects of superconducting coil design as the size and magnetic field increase is the structure necessary to maintain the mechanical integrity of the windings.

Appendix B presents the analysis of the structural require-

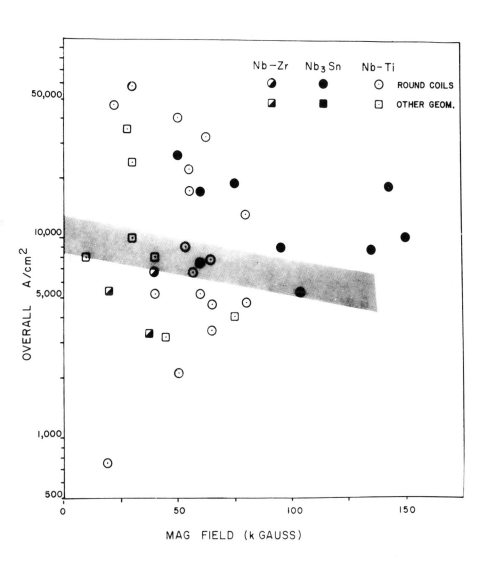

Figure 11. Current density vs. magnetic field for several superconducting coils.

ments for a long thin solenoid. The main points that are made are:

1. A "magnetic pressure" equal to $B^2/2\mu_0$ acts on the windings. (This pressure is more like stretched rubber bands than like a hydrostatic pressure). In general this pressure acts to expand the coil system as a whole and to compress or "pinch" the conductor bundles that carry current in the same direction.

2. The total structural mass of a coil system is given by:[21]

$$M = K \frac{\gamma}{\sigma} E_m, \qquad (5)$$

where
- M = structural mass,
- γ = density of the structural material,
- σ = working stress of the structure,
- E_m = total magnetic energy stored.

The constant K depends on the particular coil geometry and structure and is shown to be equal to 3 for the case of a long thin solenoid. It is a measure of the effectiveness of a structure to store magnetic energy.

In general, simple tension structures result in lower values of K than do structures utilizing compression or bending members.

Noncircular coil geometries have more complex structures and consequently have higher values of K.

For geometrically similar windings, structural weight increases as the square of the magnetic field and as the cube of the physical size, so that for large high field strength coils structural considerations are one of the major, if not the most important, aspects of the design.

STATE OF THE ART

The state of the art of superconducting magnet size versus the magnetic field at the windings[22-33] is shown in Figure 12.

The coils have been plotted according to the superconductor and geometric shape. The internal diameter is plotted for circular or round coils. Noncircular coils are plotted at a diameter equal to that of a circle of area equal to the flux area in the useful region.

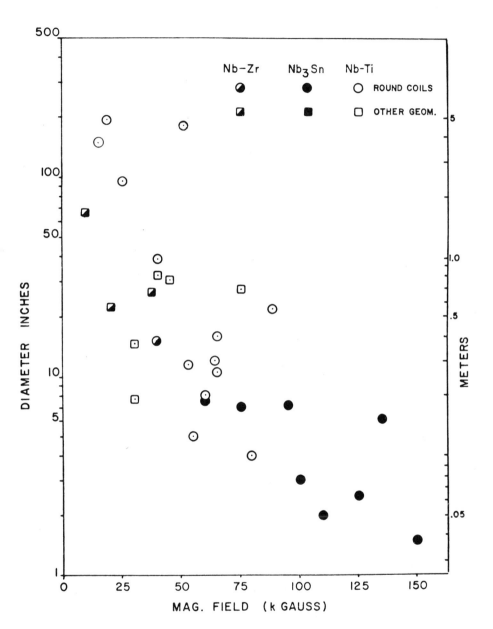

Figure 12. Effective diameter vs. magnetic field for superconducting coils.

Examination of the figure shows that the alloy superconductors are used to about 90 kilogauss. Above this Nb_3Sn is used. Although Nb_3Sn is capable of use below these fields, there are relatively few sizeable Nb_3Sn coils at lower fields. This is because Nb_3Sn conductors are relatively more expensive than NbTi conductors, so that unless current density is of prime importance NbTi is likely to be the first choice.

The energy storage capability of superconducting coils is shown in Figure 13 which depicts the specific weight in kg/Mj versus the energy stored in Megajoules. The data points are for magnets designed for steady operation.

The data for weight includes the windings and structure, but not the cryogenic container or refrigerator, so the plot represents that of a structurally sound superconducting energy storage unit.

No coils were plotted that had either iron return paths or other heavy components that would not be used if weight were an important parameter. Also, no points were plotted where it was felt that another component added unnecessarily to the magnet weight.

The high current density of Nb_3Sn coils shows their low specific weight at energies from 0.01 to 1 Mjoule. The circular coils made of alloy superconductors are heavier; however, the same downward slope of specific weight versus energy is exhibited for this class of coils. The apparent lower specific weight of Nb_3Sn has been used (with few exceptions) primarily in small magnets. These usually are not encumbered by structural problems or problems of dissipating a large magnetic energy during the quench.

All of the data points in Figure 13 correspond to either Nb_3Sn coils or to coils of relatively low current density using steady state stability criteria. This is so because all the high current density magnets have so far been restricted to relatively low energies. Further development of larger high current density coils will undoubtedly show lower specific weights for coils made with alloy superconductors.

For comparison, high performance capacitors have a specific weight of 10,000 kg/Mj, while a battery has a specific weight of 2 kg/Mj.

The data points in Figure 13 for each class of coils show a decreasing specific weight with increasing energy. As has already been discussed, the structural limitations are a lower limit to the weight per unit energy stored. This limit is not evident from the data points and extrapolation to higher energies should be done keeping in mind the structural requirements.

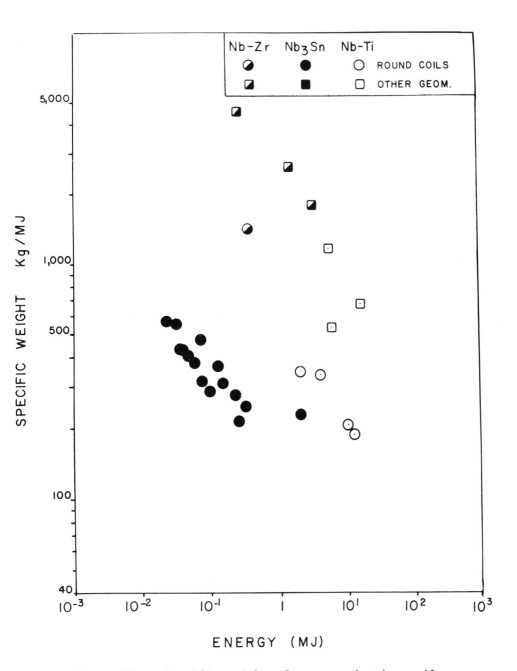

Figure 13. Specific weight of superconducting coils.

SPECIFIC APPLICATIONS

To illustrate the uses superconducting magnets have been put to, four coils will be discussed. The four coils are as follows:

1. A typical Magnetic Corporation of America laboratory research magnet.

2. The coils for the Argonne National Laboratory 3.6 m bubble chamber for high energy physics research.[32]

3. The Lawrence Radiation Laboratory Baseball II coil system for fusion research.[25]

4. An advanced transverse field magnet for magnetohydrodynamic power generation being built by Magnetic Corporation of America. (Under the sponsorship fo the United States Air Force Aero Propulsion Laboratory, Wright Patterson Air Force Base.

A summary of the main characteristics of these four coil systems is presented in Table III.

The laboratory coil is shown in Figure 14. It is designed for higher than average uniformity and makes use of correcting windings at both ends. It is wound with multifilament small diameter round wire. The operating overall current density for this coil is 17,000 A/cm^2. (Values up to twice this value can be achieved by using a conductor with a lower ratio of copper to superconductor.) Coils of this type are tightly wound without any provision for internal cooling passages. Charging times are generally short, being no longer than a few minutes. Current densities are generally high and stability is determined more by the particular conductor used than any other construction detail. No quench protection is required for coils of this general size since the magnet energy is generally no higher than several tens of kilojoules.

The coils for the Argonne National Laboratory bubble chamber are shown under construction in Figure 15.

The superconducting coils and their iron return path provide the 1.8T for the 3.6 m diameter bubble chamber at the Argonne National Laboratory which is used in studying the interactions of high energy particles. The bubble chamber itself makes the particle tracks visible, while the magnetic field curves the path of charged particles, thus allowing them to be identified.

These are currently the largest operating superconducting windings both in terms of size as well as magnetic energy stored. Two sets of coils (one set is shown in Figure 15) are mounted vertically one above the other. The conductor used operates at a

Table III. Characteristics of Coils for Typical Applications

	Laboratory Magnet	ANL Bubble Chamber Coils	Baseball II	Advanced MHD Magnet
Type of Field	end corrected solenoid	iron clad solenoid	minimum B	transverse
Magnetic Field (T)	6.5	1.8	2.0	5.0
Volume	5 cm dia. 15 cm long	4.8 m dia. 3.05 m long	1.2 m sphere	0.21 m dia. 0.90 m long
Uniformity over Useful Volume (%)	.01 over 2.5 cm sphere	2.5		5
Magnetic Field (T) at the Windings	6.5	1.8	7.5	7
Conductor:				
Superconductor	Nb-Ti	Nb-Ti	Nb-Ti	Nb-Ti
Copper/Superc. ratio	2:1	20:1	4:1	1.8:1
Dimensions	.05 mm dia.	50 x 2.5 mm	6.35 x 6.35 mm	1.62 x 1.62 mm
No. of Strands	121	6	24	180
No. Twists/cm	4	0	0	0.5
Operating Current	50	2200	2400	520
Conductor j (A/cm^2)	24,000	1700	5960	20,000
Overall j (A/cm^2)	17,000	775	4000	15,000
Conductor Weight (kg)	15	45,000	4550	224
Structure Weight (kg)	no separate struct.	no separate struct.	6800	146
Magnetic Energy (MJ)	.02	80	17	1.6

Figure 14. Typical laboratory superconducting magnet.

Figure 15. Picture of one-half of the 16 foot (4.8 m) dia. 18 kg coil system under construction at the Argonne National Laboratory. These coils are being used in the bubble chamber.

relatively low current density of 1700 A/cm^2, and like most large magnet systems is designed using static stability design criteria. The conductor contains six strands of Nb-Ti and was manufactured before the benefits of twisting were understood. The rectangular conductor is wound into pancakes which were assembled with spacers that form cooling passages between pancakes. In this way, each conductor is cooled at both edges along its whole length.

Other coils of this general size have been designed to allow helium in between the conductors so that additional heat transfer area is available, which results in higher allowable current densities. Hollow conductors have also been used in large coils with supercritical helium circulating within the conductor itself.[33]

Another variation on the construction in similar coils has been the use of reinforcing strips of stainless steel of the same width as the conductor which are wound into the pancakes simultaneously with the conductor itself.

The Baseball II coil system shown in Figure 16 provides a so-called minimum B magnetic field configuration. In this configuration the magnetic field is minimum at the center of the coil system, about 2T, and increases in magnitude in all directions away from the center. This particular geometry of magnetic field is useful in thermonuclear fusion research and serves to confine the energetic particles near the center of the system to allow them to interact. The name Baseball comes from the similarity of the coil windings to the seam on a baseball.

The conductor used for the Baseball II magnet is a Nb-Ti composite, 6.35 mm square, containing 24 strands of superconductor. The copper to superconductor ratio is 4 to 1, and the strands are untwisted. The design is based on statically stable design criteria with the conductor cooled with helium along its length. Although the magnetic field in the center of the coil bore is 2T, it increases to 7.5T at the conductor surface. The operating overall current density is 4000 A/cm^2.

The importance of the magnetic forces is evident from Figure 16 which shows the large structural components which completely enclose and contain the windings.

Figure 17 shows a schematic of a transverse field magnet for research in magnetohydrodynamic power generation. An electrically conducting gas flows through the magnet bore and generates electricity directly from the interaction of the gas with the magnetic field. This magnet makes use of the conductor shown in Figure 5 which is typical of the twisted multifilament composites. The magnet has no internal cooling passages and has an overall current density higher than would be possible using static stability criteria.

SUPERCONDUCTING COILS

Figure 16. Minimum B Configuration Baseball II superconducting magnet built for controlled thermonuclear reaction research at the Lawrence Radiation Laboratory. The coils have a design field of 75 kilogauss at the winding and a useful volume of approximately 1 meter in diameter.

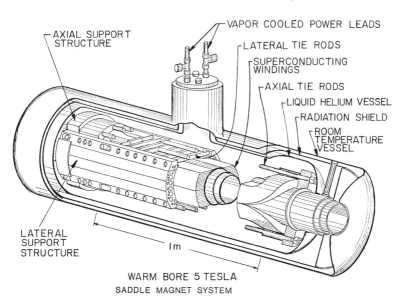

Figure 17. Schematic diagram of a transverse field superconducting magnet in a cryogenic container with a room temperature bore.

APPENDIX A - STEADY STATE STABILITY

Introduction

The behavior of a composite conductor is determined by the characteristics of the superconductor, the substrate which surrounds it, and the characteristics of the cooling medium used to carry away the heat.

This treatise considers the general case of a composite conductor operating under steady conditions, taking into account the temperature drop within the superconductor itself, a thermal interface resistance between the superconductor and the substrate, and the cooling of the conductor by a coolant.

The emphasis is on the physics of the phenomenon rather than the derivation of a set of equations that can accurately predict the terminal characteristics of a particular composite conductor.

Formulation of Equations

Consider the general case of a conductor composed of superconductor (in many strands) and a cooled stabilizing substrate which completely surrounds each strand of superconductor.

The following assumptions will be made:

1. All the superconducting strands have equal properties and have the same shape and size.

2. There are no thermal gradients in the stabilizing substrate.

3. The heat generated within a superconducting strand is transferred to the substrate through a surface thermal contact resistance which is assumed to be uniform throughout.

4. There are no thermal gradients along the conductor, so that we can deal with heat flow transverse to the conductor length.

5. The current density at any point within each superconducting strand is assumed to depend only on the local temperature. Also the current density is assumed to decrease to zero linearly with increasing temperature:

$$j_s = j_{ch} \left[1 - \frac{T_s - T_b}{T_{ch} - T_b} \right], \qquad (A-1)$$

where j_s = local superconductor current density,

j_{ch} = local superconductor current density at bath temperature,

T_s = local superconductor temperature,

T_b = bath or coolant temperature,

T_{ch} = temperature at which the current density goes to zero. (In the presence of whatever magnetic field exists.)

The temperature distribution within each superconducting strand is determined by heat conduction toward the superconductor-substrate interface and by the heat generation within the superconductor when a voltage exists across it. Without specifying the detail shape of the superconducting strands, we can write down the equation for their temperature distribution:

$$k\nabla^2 T_s = v \cdot j_s , \qquad (A-2)$$

where k = thermal conductivity (assumed constant),

V = voltage per unit length of conductor.

The right hand side of Eq. (A-2) represents the heat generated per unit volume within the superconductor itself.

At the surface of the superconducting strand the temperature is assumed to be uniform and equal to T_w.

The total current carried by all the superconducting strands is:

$$I_s = n \int j_s \, dA_s , \qquad (A-3)$$

where n is number of superconducting strands.

All the heat generated within a superconducting strand flows through the thermal contact resistance at the superconductor-substrate interface into the substrate:

$$v \cdot \left(\frac{I_s}{n}\right) = h_i P_i (T_w - T_{sub}) , \qquad (A-4)$$

where h_i = superconductor-substrate interface heat transfer coefficient,

P_i = perimeter of one strand of superconductor in contact with the substrate,

T_w = surface temperature of the superconductor.

All the heat generated within the conductor passes from the substrate to the coolant:

$$v \cdot I = hP(T_{sub} - T_b) \ , \qquad (A-5)$$

where I = total current in the conductor,

h = heat transfer coefficient at the substrate-coolant interface,

P = perimeter exposed to the coolant.

The voltage per unit length of conductor is determined by the current that flows in the normal substrate:

$$v = \frac{\rho}{A}(I - I_s) \ , \qquad (A-6)$$

where ρ = resistivity of the substrate,

A = cross-sectional area of the substrate.

It is very informative to put the equations into dimensionless form:

$$\frac{j_s}{j_{ch}} = 1 - \frac{T_s - T_b}{T_{ch} - T_b} \ , \qquad (A-7)$$

$$\frac{kA(T_{ch} - T_b)}{\rho I_{ch} j_{ch} r_w^2} \nabla'^2 \left(\frac{T_s - T_b}{T_{ch} - T_b} \right) = \left(\frac{vA}{\rho I_{ch}} \right) \left(\frac{j_s}{j_{ch}} \right) \ , \qquad (A-8)$$

where ∇' has been non-dimensionalized with respect to the half width or half thickness of a superconducting strand r_w,

$$\frac{I_s}{I_{ch}} = \int \left(\frac{j_s}{j_{ch}} \right) \frac{dA_s}{A_s} \ , \qquad (A-9)$$

SUPERCONDUCTING COILS

$$\left(\frac{vA}{\rho I_{ch}}\right)\left(\frac{I_s}{I_{ch}}\right) = \left(\frac{nh_i P_i A (T_{ch}-T_b)}{\rho I_{ch}^2}\right)\left(\frac{T_w - T_{sub}}{T_{ch}-T_b}\right), \quad (A-10)$$

$$\left(\frac{vA}{\rho I_{ch}}\right)\left(\frac{I}{I_{ch}}\right) = \frac{hPA (T_{ch}-T_b)}{\rho I_{ch}^2} \left(\frac{T_{sub}-T_w}{T_{ch}-T_b}\right), \quad (A-11)$$

and

$$\left(\frac{vA}{\rho I_{ch}}\right) = \frac{I}{I_{ch}} - \frac{I_s}{I_{ch}} . \quad (A-12)$$

In the above equations

I_{ch} = critical current = $j_{ch} A_s$,

A_s = total superconductor cross section,

r_w = half thickness or radius of superconducting strand.

Let us now define the following variables:

$$\mathcal{J} = j_s/j_{ch} , \quad (A-13)$$

$$\theta = \frac{T-T_b}{T_{ch}-T_b} \quad \text{(with appropriate subscripts, s or w),} \quad (A-14)$$

$$V = \frac{vA}{\rho I_{ch}} , \quad (A-15)$$

$$\tau = I/I_{ch} , \quad (A-16)$$

$$\alpha = \frac{\rho I_{ch}^2}{hPA (T_{ch} - T_b)} , \quad (A-17)$$

$$\alpha_i = \frac{\rho I_{ch}^2}{nh_i P_i A(T_{ch}-T_b)} , \quad (A-18)$$

$$r_{cs} = \sqrt{\frac{k(T_{ch}-T_b)A}{\rho j_{ch} I_{ch}}} = \frac{A}{I_{ch}} \sqrt{\frac{k(T_{ch}-T_b)}{\rho} \frac{A_s}{A}} . \quad (A-19)$$

With these variables the equations reduce to:

$$\dot{\mathscr{J}} = 1 - \theta_s, \tag{A-20}$$

$$\left(\frac{r_{cs}}{r_w}\right)^2 \nabla'^2 \theta_s = V \dot{\mathscr{J}}, \tag{A-21}$$

$$\left(\frac{I_s}{I_{ch}}\right) = \int \dot{\mathscr{J}} \, \frac{dA_s}{A_s}, \tag{A-22}$$

$$V\left(\frac{I_s}{I_{ch}}\right) = \frac{\theta_w - \theta_{sub}}{\alpha_i}, \tag{A-23}$$

$$V\tau = \frac{\theta_{sub}}{\alpha}, \tag{A-24}$$

$$V = \tau - \left(\frac{I_s}{I_{ch}}\right). \tag{A-25}$$

Eliminating and rearranging the above equations, we arrive at:

$$V = \tau - 1 + \alpha V\tau + \alpha_i V(\tau - V) + \int (\theta_s - \theta_w) \frac{dA_s}{A_s}. \tag{A-26}$$

The quantity $\theta_s - \theta_w$ is the dimensionless temperature rise in the superconducting strands.

Making use of Equations A-20 and A-21, it can readily be shown that:

$$\nabla'^2[(\theta_s - \theta_w) - (1 - \theta_w)] + V\left(\frac{r_w}{r_{cs}}\right)^2 [\theta_s - \theta_w - (1-\theta_w)] = 0, \tag{A-27}$$

which has a solution of the form:

$$\theta_s - \theta_w = (1 - \theta_w) + F\left[\left(\frac{r}{r_{cs}}\right)\sqrt{V}\right], \tag{A-28}$$

where $F[\ ]$ denotes a functional relationship. At $r = r_w$, $\theta_s = \theta_w$

and the quantity $1 - \theta_w$ can be eliminated:

$$\theta_s - \theta_w = F\left[\left(\frac{r}{r_{cs}}\right)\sqrt{V}\right] - F\left[\left(\frac{r_w}{r_{cs}}\right)\sqrt{V}\right]. \quad (A-29)$$

The quantity under the integral sign in Eq. (A-26) is simply the average value of $\theta_s - \theta_w$ over the superconductor area, which means integration over r, so:

$$\int (\theta_s - \theta_w) \frac{dA_s}{A_s} = G\left[\left(\frac{r_w}{r_{cs}}\right)\sqrt{V}\right], \quad (A-30)$$

where G [ ] represents a functional relationship. Substitution into Eq. (A-26) yields the final relationship:

$$V = \tau - 1 + \alpha V\tau + \alpha_i V(\tau - V) + G\left[\left(\frac{r_w}{r_{cs}}\right)\sqrt{V}\right]. \quad (A-31)$$

Given the quantities α, α_i, and r_w/r_{cs}, and the geometry of the strands, so that the functional relationship G is defined, the curve of V versus τ is completely determined.

Discussion

Equation (A-31) above shows very graphically the effects of heat transfer to the coolant, the interface thermal resistance, and the conductor size. Very simply the equation

$$V = \tau - 1 \quad (\tau > 1) \quad (A-32)$$

is the dimensionless voltage per unit length for a superconductor and a substrate; all current above the critical current ($\tau = 1$) flows in the normal substrate and produces a voltage proportional to the excess current over the critical value. As yet, no account has been taken of the heating of the conductor as a whole, which is the product of the voltage times the current.

If the next term is added to Equation (A-32):

$$V = \tau - 1 + \alpha V\tau \quad (\tau > 1). \quad (A-33)$$

The additional term is indeed proportional to the dimensionless voltage times the dimensionless current $V\tau$, and the proportionality constant

$$\alpha = \frac{\rho I_{ch}^2}{hPA(T_{ch} - T_b)}$$

is simply the temperature rise with all the current in the substrate divided by the temperature rise for the current density to go to zero in the superconductor.

The term $\alpha V\tau$ can be thought of as an additional voltage which appears across the superconductor due to imperfect cooling by the coolant. The quantity α is a measure of the cooling effectiveness.

The next term to be added is $\alpha_i V(\tau - V)$. Referring to Equation (A-25), this term can be written as $\alpha_i V(I_s/I_{ch})$. In this form it is analogous to the $\alpha V\tau$ term, with the difference that the interface perimeter and heat transfer coefficient are used, and that the temperature drop across the interface is proportional to the voltage times the superconductor current rather than the total current, so that the quantity τ is replaced by I_s/I_{ch}.

The constant of proportionality, α_i, is analogous to the quantity α except that the interface heat transfer coefficient, h_i, and total perimeter of contact, nP_i, are used.

The last term in Eq. (A-31) contains the quantity $(r_w/r_{cs})\sqrt{V}$. It is evident that if $V = 0$ there is no heating in the superconductor, and therefore $G(0) = 0$.

The effect of superconductor size and its interaction with other quantities is best revealed by examining the slope of the voltage-current characteristic at the critical current ($\tau = 1$) when voltage first begins to appear ($V = 0+$).

Taking the derivative of Eq. (A-31) with respect to V:

$$1 = \frac{d\tau}{dV} + \alpha \left[\tau + V\frac{d\tau}{dV}\right] + \alpha_i \left[(\tau-V) + V\left(\frac{d\tau}{dV} - 1\right)\right] + \frac{dG}{dV}. \quad \text{(A-34)}$$

At $\tau = 1$ and $V = 0$ we can solve for $d\tau/dV$:

$$\frac{d\tau}{dV} = 1 - \alpha - \alpha_i - \frac{dG}{dV}. \quad \text{(A-35)}$$

The derivative dG/dV is evaluated as follows:

$$\frac{dG[u]}{dV} = \frac{dG}{du} \cdot \frac{du}{dV} = \frac{1}{2}\frac{du^2}{dV} \cdot \frac{dG}{u\,du} = \frac{1}{2}\left(\frac{r_w}{r_{cs}}\right)^2 \frac{dG}{u\,du}, \quad \text{(A-36)}$$

SUPERCONDUCTING COILS

where

$$u = \frac{r_w}{r_{cs}} \sqrt{V} .$$

Thus

$$\left.\frac{d\tau}{dV}\right|_{\tau=1} = 1 - \alpha - \alpha_i - \frac{1}{2}\left(\frac{r_w}{r_{cs}}\right)^2 \left[\frac{dG}{u\,du}\right]_u . \quad (A-37)$$

The quantity $dG/u\,du$ is a function of u only, and as $u \to 0$ it takes on a numerical value determined by the shape of the superconductor.

Taking the inverse of Equation (A-37) we have:

$$\left.\frac{dV}{d\tau}\right|_{\tau=1} = \frac{1}{1-\alpha-\alpha_i - \frac{1}{2}\left(\frac{r_w}{r_{cs}}\right)^2 \left[\frac{dG}{u\,du}\right]_{u=0}} . \quad (A-38)$$

The general condition for stability against small voltage excursions is to have $dV/d\tau$ finite, which requires the denominator to be non-zero.

This leads to the general stability requirement that

$$\alpha + \alpha_i + \frac{1}{2}\left(\frac{r_w}{r_{cs}}\right)^2 \left[\frac{dG}{u\,du}\right]_{u=0} < 1 \quad (A-39)$$

for stability against small disturbances at the critical current. Stability against large disturbances requires detailed knowledge of the complete V-τ curve.

APPENDIX B - STRUCTURAL REQUIREMENTS

The large variety of coil sizes and shapes required for different applications makes it important to have a basis for structural design. In order to arrive at this basis, the structural weight for a simple case will be calculated - namely an infinitely long solenoid with windings that are thus compared with the diameter, as shown in Figure (B-1)

The magnetic field generated inside the solenoid is:

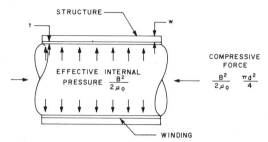

Figure B-1. Forces acting on a long thin solenoid.

$$B = \mu_0 j t, \tag{B-1}$$

μ_0 = permeability of free space,

j = current density in the windings,

t = winding thickness.

The magnetic field outside the coil is zero for a long solenoid, so that the total force per unit area pushing outward radially is simply the average magnetic field, $B/2$, times the current per unit axial length, jt. Making use of the expression for the internal magnetic field, we arrive at the following result:

$$\text{Radial Force/Area} = (jt)\left(\frac{B}{2}\right) = \frac{B^2}{2\mu_0}. \tag{B-2}$$

This radial force per unit area is simply the "magnetic pressure" - well known to plasma physicists.

In addition to the magnetic pressure acting outward radially, there is a force generated at the coil ends which compresses the windings axially. This axial compression is a result of the radial component of magnetic field which exists only near the ends of the long solenoid.

In terms of the radial field, B_r, near the ends, the expression for the total axial compressive force is:

$$F_a = \int j B_r t (\pi d) dz, \tag{B-3}$$

where

d = diameter of the solenoid,

z = axial distance,

and the integration is carried out over one end of the solenoid only. (It has been assumed in the above expression that B_r does not vary

SUPERCONDUCTING COILS

within the windings of thickness t, a good assumption for ratios of t/d that are small.)

The integral can be rewritten as follows:

$$F_a = jt \int \pi dB_r dz. \tag{B-4}$$

The quantity under the integral sign is simply the magnetic flux which crosses the windings in a radial direction. It can be shown for a long solenoid that one half of the total flux crosses the windings, the other half of the flux leaves through the bore of the solenoid at the ends. Under these circumstances

$$\text{TOTAL FLUX} = \frac{\pi d^2}{4} \cdot B = 2 \int \pi dB_r dz. \tag{B-5}$$

Making this substitution yields the expression for the axial force as

$$F_a = \left(\frac{\pi d^2}{4}\right)\left(\frac{B}{2}\right) jt = \frac{\pi d^2}{4}\left(\frac{B^2}{2\mu_0}\right). \tag{B-6}$$

This can be thought of as a magnetic pressure, $B^2/2\mu_0$, acting axially across the bore of the solenoid.

The following table shows the equivalent value of the magnetic pressure in N/m^2 and lb/in^2 for various values of magnet field in Tesla:

Magnetic Field	Magnetic Pressure	
Tesla	lb/in^2	N/m^2
20	24,000	1.59 x 10^8
10	6,000	3.96 x 10^7
5	1,500	0.99 x 10^7
2	240	1.59 x 10^6
1	60	3.96 x 10^5
0.5	15	0.99 x 10^5
0.2	2.4	1.59 x 10^4
0.1	0.6	3.96 x 10^3

It is very evident from the above that at high fields structural considerations are of primary importance.

Returning now to the example of the long thin solenoid, the structure required is one that supports the radial pressure as well as the axial compression.

If a single cylinder of thickness w is used as the structure, it will have the following stresses:

$$\text{hoop stress (tensile)} = \frac{d}{2w} \cdot \frac{B^2}{2\mu_0}, \tag{B-7}$$

$$\text{axial stress (compressive)} = \frac{d}{4w} \cdot \frac{B^2}{2\mu_0}. \tag{B-8}$$

For a biaxial stress situation such as this, the equivalent stress in tension, σ, is equal to the sum of the absolute magnitudes of the two stresses:

$$\sigma = \frac{3}{4} \frac{d}{w} \frac{B^2}{2\mu_0}, \tag{B-9}$$

and the thickness w is:

$$w = \frac{3}{4} \frac{d}{\sigma} \frac{B^2}{2\mu_0}. \tag{B-10}$$

The total mass of the structure of density γ for a total axial length ℓ is then:

$$M = \gamma \pi dw\ell = 3\frac{\gamma}{\sigma}\left[\frac{\pi d^2 \ell}{4} \cdot \frac{B^2}{2\mu_0}\right]. \tag{B-11}$$

The quantity in brackets is the volume of the bore times the quantity $B^2/2\mu_0$, which in addition to being the magnetic pressure is also equal to the magnetic energy density. The quantity in brackets therefore is simply the total magnetic energy, E_m, and the following relationship results:

$$\frac{M}{E_m} = 3\frac{\gamma}{\sigma}. \tag{B-12}$$

This relationship states that the mass of structure required is directly proportional to (1) the ratio of density to strength of the structural material used, as well as (2) the total magnetic energy stored in the coil.

The factor of 3 in the equation is characteristic of the particular coil and structural configuration (cylinder).

In general the relationship for structural mass takes the form

$$\frac{M}{E_m} = K \frac{\gamma}{\sigma}, \quad (B-13)$$

where the constant K is a constant that depends on the particular coil and structural configuration. The constant K can be used to choose between alternate structural configurations.

It is important to emphasize at this point that the ultimate limitation on energy storage per unit mass capability of a superconducting coil is a structural one.

REFERENCES

[1] Britton, R. B., Proc. 1968 Summer Study Supercond. Devices and Accelerators, Brookhaven National Laboratory, BNL 50155, p. 449, 1968.

[2] Stekly, Z. J. J., Journal of Applied Physics Vol. 42, No. 1, p. 65, Jan. 1971.

[3] Stekly, Z. J. J., et al., BNL 50155, p. 748.

[4] Gauster, W. F., and Hendricks, J. B., Journal of Applied Physics Vol. 39, p. 2572, 1968.

[5] Maddock, B. J., et al. Cryogenics, Vol. 9, No. 4, p. 261, August, 1969.

[6] Kremlev, M. G., Cryogenics, Vol. 7, p. 267, 1967.

[7] Chester, P. F., Rep. Prog. Phys. XXX, Part II, pp. 561-614, 1967.

[8] Hart, H. R., BNL 50155, pp. 571-600.

[9] Rutherford Laboratory Preprint, RPP/A73, November 1969.

[10] Hancox, R., Proc. I.E.E.E., 113, pp. 1221-1228, 1966.

[11] Hancox, R., I.E.E.E. Trans. Magnetics, MAG-4, pp. 486-8, 1968.

[12] Laverick, C., Argonne National Laboratory, ANL/HEP 6810, May, 1968.

[13] Smith, P. F., Wilson, M. N., Walters, C. R., and Lewin, J. D., BNL 50155, p. 913.

[14] Wittgenstein, F., et al., Cryogenics, pp. 158-164, June 1969.

[15] Benz, M. G., Proceedings of Les Champs Magnetiques Intenses Leur Production et Leurs Applications, Grenoble, Sept. 1966., Editions du Centre National de la Recherche Scientifique, Paris, 1967.

[16] Benz. M. G., General Electric Report No. 66-C-044, Feb., 1966.

[17] Fairbanks, D. F., Advances in Cryogenic Engineering, Vol. 14, p. 133, Plenum Press, New York, 1969.

[18] Stekly, Z. J. J., et al., Journal of Applied Physics, Vol 39, 6, p. 264, 1968.

[19] Stekly, Z. J. J., Advances in Cryogenic Engineering, Vol. 8, p. 585, New York, Plenum Press, 1963.

[20] Stekly, Z. J. J., Journal of Applied Physics, Vol. 42, No. 1, pp. 65-72, Jan. 1971.

[21] Levy, R. H., American Rocket Society Journal, p. 787, May 1962.

[22] Fast, R. F., et al., Journal of Applied Physics, Vol. 42, No. 1, p. 79 Jan. 1971.

[23] Rogers, J. D., et al., Journal of Applied Physics, Vol. 42, No. 1, p. 73, Jan. 1971.

[24] Lucas, E. J., et al., Advances in Cryogenic Engineering, Vol. 15, p. 167, 1969.

[25] Henning, C. D., et al., Advances in Cryogenic Engineering, Vol. 14, p. 98, 1968.

[26] Coles, W. D., Advances in Cryogenic Engineering, Vol. 13, p. 142, 1967.

[27] Stekly, Z. J. J., Electricity from MHD, International Atomic Energy Agency, Vienna, 1968.

[28] Cryogenic Engineering (Japanese). Vol. 5, No. 2, 1970. (Whole Issue).

[29] Berruyer, A., et al., Advanced Cryogenic Engineering, Vol. 15, p. 158, 1969.

[30] Proc. 1968 Summer Study Supercond. Devices and Accelerators, Brookhaven National Laboratory, BNL 50155, 1968.

[31] Montgomery, D. B., et al., Advances in Cryogenic Engineering, Vol. 14, p. 88, 1968.

[32] Purcell, J. R., Argonne National Laboratory Report ANL/HEP 6813, June 1968.

[33] Meyer, G., Maix, R., Brown-Boveri Review (Date Unknown).

PHYSICS OF SUPERCONDUCTING DEVICES

Bascom S. Deaver, Jr.

Physics Department, University of Virginia

Charlottesville, Virginia   22901

So far this course has been concerned primarily with the science of superconductivity, the various phenomena observed experimentally and the phenomenological and microscopic descriptions which generalize these phenomena within the framework of theoretical physics. Many of the lectures to follow will discuss applications or the technology of superconductivity. This lecture is intended to serve as a kind of interface between these two major topics of the course and it has two primary objectives:

1. To give a brief overview of the many applications of superconductivity

2. To discuss the physics of that subset of applications sometimes called "quantum devices", that is, those involving fluxoid quantization, flux flow and the Josephson Effect.

APPLICATIONS OF SUPERCONDUCTIVITY

The table on the following pages is an attempt to summarize the applications of superconductivity. One striking feature is the very diversity of them. They range from the very large scale--transmission lines carrying 10^6A, magnets producing fields in excess of 10^5 Gauss, stored energy greater than 10^7 joules--to the extremely small--devices with dimensions of a few microns, switching times of 10^{-11} sec., sensors capable of measuring 10^{-11} Gauss or 10^{-17} Volts and shields that potentially produce zero magnetic field. Although the science of superconductivity dates

Applications of Superconductivity[1]

Sensors

Physical Entity Sensed or Detected	Property or Device Used	Remarks
Temperature	$R(T)$, $\lambda(T)$ near T_c	Fast response, $\Delta T \sim 10^{-6}$K
	Linewidth of Josephson Oscillation	Possible fundamental thermometer in millidegree range
Magnetic Field	Meissner Effect, quantized flux; Josephson Effect	$\Delta B \sim 10^{-11}$ Gauss
Magnetic Flux	"	$\Delta\phi \sim 10^{-11}$ G cm^2
Magnetic Field Gradients	"	$dB/dx \sim 10^{-9}$ G/cm
Current	"	$\Delta i \sim 10^{-9}$A at very low impedance
Emf	"	$\Delta V \sim 10^{-15}$V
Position	Change in inductance of coil or meander line with position of sc core or plane	$\Delta x \sim 10^{-13}$cm
Acceleration	Magnetic support and magnetometer read out	10^{-12}g
Angular Position	Magnetic support, London Moment	Probably seconds of arc
Angular Velocity	London Moment	
Electromagnetic	Bolometer using $R(T)$ near T_c; Josephson Effect	10^{-14} watts, rf through infrared
Charged Particles	Resistive transition of current biased film	Spatially localized detection
	Josephson Junction ion chamber	
Phonons	Josephson Junction	Detects single phonon

PHYSICS OF SUPERCONDUCTING DEVICES

Applications of Superconductivity (con't)

	Property or Device Used	Remarks
Signal Processors		
Amplifiers	Cryotron, parametric devices	
Detectors, Mixers Harmonic Generators	Josephson Junctions, weak links, thin films	
Delay Lines	R = 0, kinetic inductance	
Digital Devices	Quantized flux storage, switching of junctions and weak links	Fast, inherently digital
Magnets	R = 0, high pinning force	Stable fields in persistent mode, 150 kG, large volume
Magnetic Shields	R = 0, fluxoid quantization	Near perfect electromagnetic shielding, 10^{-6}-10^{-8} G in volume $\sim 1 m^3$, Potentially B = 0
Magnetocardiography	Magnetometer, gradiometer	
Magnetic Susceptometry	Magnetometer, shield, magnet	
Spectrometry	Josephson Effect	
High Q Microwave Cavities		Unloaded Q $\sim 10^{11}$
Linear Accelerators		cw operation, $\Delta E/E \sim 10^{-4}$
Voltage Standard	Josephson Effect	~ 1 ppm
Temperature Standards	Fixed points at T_c of materials with sharp transitions	
Thermometry	Magnetometer, magnet	Using Susceptibility change of paramagnetic material
Heat Switches	Change in thermal conductivity between normal and sc state	
Energy Transmission and Storage		10^6 A, > 10^7 joules
Motors and Generators		3000 hp, 25 M watt
Mechanical Support	Persistent magnets	Trains, structures

from 1911, the technology is largely a product of the last decade and is currently expanding at a great pace. Magnets must constitute the largest commercial item at present although within the last two years several companies have begun to market superconducting magnetometers and other quantum devices.

Most of these applications will be described in detail in subsequent lectures, so I will only allude to them briefly here. As sensors of magnetic field or magnetic flux, superconducting devices have a unique place because of their extreme sensitivity, fast response and small size. Magnetometers have been developed using the Meissner Effect, fluxoid quantization and the Josephson Effect. Any of these can be converted to an ammeter or voltmeter with advantages of very low impedance or high sensitivity. Various gyroscopic devices use magnetically supported superconducting rotors. One unique indication of rotation is the London moment, the magnetization of a rotating superconductor along its axis of rotation. It can be sensed by a superconducting magnetometer and used to determine the orientation of the rotor.

Because of the Josephson Effect, superconductors can be used to generate and to detect electromagnetic radiation from radio frequencies through infrared. A great variety of applications is possible using superconductors as mixers, harmonic generators and parametric devices over a very wide frequency range. A topic of great current interest is the storage of digital information using Josephson Junctions or other weakly linked superconductors and the possibility of using the small size and extremely high switching speeds of these devices for large scale memory and logic circuits.

Superconducting microwave cavities with Q's greater than 10^{11} have been developed. These are potentially very useful for frequency control and are already being used for an electron linear accelerator capable of cw operation and very good energy resolution.

References for all of these applications are readily obtained from a literature survey issued quarterly as a joint publication of NBS and ONR[1].

PHYSICS OF QUANTUM DEVICES

Fluxoid Quantization

Now let us take a closer look at some devices that make specific use of the macroscopic quantum properties of superconductors. As you know, a superconductor can be described by a macroscopic wave function

$$\Psi = \psi e^{i\chi} \tag{1}$$

where $|\Psi|^2 = n_s$ the density of the superfluid. Using the velocity operator $V = -(1/m^*)(i\hbar\nabla + e^*\bar{A})$, we can calculate the superfluid velocity v_s and obtain the supercurrent

$$\bar{j}_s = n_s e^* \bar{v}_s = \frac{n_s e^*}{m^*}(\hbar\nabla\chi - e^*\bar{A}) \quad . \tag{2}$$

From this result we can express $\nabla\chi$ in terms of the electrodynamic variables.

$$\nabla\chi = \frac{e^*}{\hbar}\left(\frac{m^*}{n_s e^{*2}}\bar{j}_s + \bar{A}\right) \quad . \tag{3}$$

In order that Ψ be singlevalued we demand that for any closed path lying in a superconductor $\oint \nabla\chi \cdot d\bar{s} = n2\pi$, where n is an integer. Then we have the result that

$$\oint \frac{m^*}{n_s e^{*2}}\bar{j}_s \cdot d\bar{s} + \oint \bar{A} \cdot d\bar{s} = n\Phi_o \quad . \tag{4}$$

This is the quantity that London called the fluxoid and $\Phi_o = h/2e$ is the flux quantum.

Now in particular consider a superconducting cylinder (Fig. 1) with wall thickness w much larger than the penetration depth λ. For a path within the walls the supercurrent can be arbitrarily small and the first integral in Eq. (4) can be neglected. Then $\oint \bar{A} \cdot d\bar{s} = \phi(i) + \phi_x = n\Phi_o$. That is, the total flux, consisting of the flux $\phi(i)$ produced by current in the ring and the flux ϕ_x produced by external sources, is quantized in units of Φ_o.

If the cylinder is cooled through its transition temperature in an applied field, a current will flow in the cylinder to make the resulting total flux quantized. If the field in which the cylinder was cooled below T_c is sufficiently small, the lowest energy state is for n = 0 in

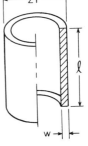

Fig. 1

Eq. (4) and a current will flow to produce exactly zero flux inside the cylinder. This property is being used to try to achieve a zero magnetic field region.

Also note that the current that flows in the cylinder is a measure of the enclosed flux and that using a superconducting ammeter to measure this current, we can determine the average field in the cylinder without access to the volume enclosed and without moving the cylinder.

In the list of applications I referred to the London moment, the magnetization of a rotating superconductor[2]. From Eq. (4) we can see that if a superconducting rod with $n = 0$ is rotated at angular velocity ω, so that within the bulk $v_s = r\omega$, the constancy of the fluxoid requires that there be a uniform field B along the axis of rotation and

$$B = \left(\frac{2m}{e}\right) \omega \sim 10^{-11} \omega \text{ Tesla} . \qquad (5)$$

In most of the rod the electrons and lattice ions rotate at the same speed; however, at the surface the superelectrons lag a little giving a field that is uniform in the interior of the rod.

There are a number of descriptions and graphic ways of thinking about superconductors that make the interpretation of some devices more intuitive. One of these is the description of the macroscopic wave function shown in Fig. 2 where the radial distance from the axis to the helix represents the amplitude of the wave function or some measure of the density of superelectrons, the angle θ is the phase of the wave function, and x is a position coordinate. For a uniform superconducting ring with $n = 1$, the amplitude is constant and the helix makes one complete turn along a path passing around the ring one time.

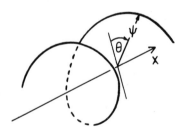

Fig. 2

Representation of the macroscopic wave function.

Now consider a cylinder with walls much thinner than the penetration depth. The current can be assumed to be uniformly distributed across the wall thickness and Eq. (4) can be written as

$$\frac{m^*}{n_s e^{*2}} \left(\frac{i}{w\ell}\right) 2\pi r + \phi(i) + \phi_x = n\Phi_o , \qquad (6)$$

where again the total flux is written in terms of the flux $\phi(i)$ produced by the supercurrent i and ϕ_x that from external sources. Further, if L is the inductance of the cylinder, $\phi(i) = Li$. Using the fact that $\lambda^2 = m^*/\mu_o n_s e^{*2}$ and that $L = \mu_o(\pi r^2/\ell)$ and if $\phi_x = 0$, we find for the total flux trapped in the cylinder[3]

$$\phi = n\Phi_o \left(1 + \frac{2\lambda^2}{wr}\right)^{-1} . \qquad (7)$$

Since λ varies with temperature and becomes large near T_c, this result can be tested by measuring ϕ as a function of temperature for a cylinder containing trapped flux. The results of such a measurement for a cylinder with n = 1 are shown in Fig. 3 where the points are measured and the curve is the calculated flux[4]. These data show that for such a thin cylinder an appreciable fraction of the phase is being contributed by the kinetic term in the fluxoid expression Eq. (4) and less by the magnetic term. (Note that it is very convenient to swap freely between thinking in terms of current and flux to thinking in terms of phase of the wave function.)

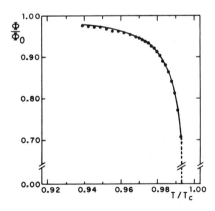

Fig. 3
Trapped flux as a function of temperature.

An interesting and sometimes useful way of thinking about the first term in Eq. (4) is in terms of a kinetic inductance, L_k. We can write Eq. (6) as

$$iL_k + iL + \phi_x = n\phi_o , \qquad (8)$$

and then identify

$$L_k = \frac{m^*}{n_s e^{*2}} \left(\frac{2\pi r}{w \ell}\right) . \qquad (9)$$

We can see what this quantity is physically if we realize that when a current flows in any conductor there are two mechanisms for storing energy. One is in the magnetic field

$$E_m = \int \frac{1}{8\pi} H^2 dV = \frac{1}{2} L i^2 \qquad (10)$$

which is one way of defining the ordinary inductance L. The second is the energy stored in the kinetic energy of the current carriers

$$E_k = \frac{1}{2} N m v^2 = \frac{1}{2} L_k i^2 , \qquad (11)$$

where N is the total number of carriers. Since $j_s = n_s e^* v_s$, Eq. (11) gives for L_k the same result as Eq. (9).

Little first discussed this idea and proposed an interesting device that he called a superinductor, a very narrow, thin strip of superconductor ($\sim$ 10μ wide and 50 A thick) for which the kinetic inductance can be 10-100 times larger than the magnetic inductance[5]. This is a useful feature for microcircuits. Further, the value of L_k can be varied with temperature and current since n_s is a function of these quantities. In fact from Eq. (8), $L_k = (n\phi_o - iL)/i$, thus the data in Fig. 3 can also be interpreted as a measure of the kinetic inductance, which in this case reaches about $\frac{1}{2}$ L near T_c before the abrupt transition to the n = 0 state.

Weakly Linked Superconductors

The important properties of weakly linked superconductors were discovered by Brian Josephson who from his calculations predicted several phenomena which were shortly after observed experimentally and now are commonly lumped under the name "Josephson Effect"[6]. His results are frequently summarized by the following two equations:

$$j = j_c \sin \Theta \qquad \text{where } \Theta = \chi_2 - \chi_1 \qquad (12)$$

$$\dot{\Theta} = \frac{2e}{\hbar} V \ . \qquad (13)$$

The first equation relates the current to the difference in phase between the macroscopic wave functions on each side of the weak link. The second relates the rate of change of the phase difference to the voltage across the weak link. (Later I will mention briefly the spatial variation of the phase difference which is related to the magnetic field.) The fundamental basis for these results will be presented by Professor Bloch in the next lecture.

As a vehicle for an experimental approach to the physics of weakly linked superconductors, let us focus again on a superconducting ring but this time with a very narrow neck in it as shown in Fig. 4. If we evaluate Eq. (4) around the path C and if the ring is thick with respect to λ everywhere except at the weak link, the only contribution to the first integral will be across the link and for a very short link the second will be very nearly the total flux in the ring. We can again write Eq. (4) as

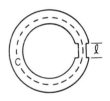

Fig. 4

$$iL_k + iL + \phi_x = n\Phi_o \qquad (14)$$

where now $L_k = m^*\ell/n_s e^{*2}\sigma$ with ℓ the length of the link and σ its cross sectional area. As we said earlier, n_s depends not only on temperature but also on current, so the phase shift

$$\Theta = \frac{2\pi}{\Phi_o} \ i \ L_k(i) \qquad (15)$$

depends on the current through the link. For a link in which the current can be considered uniformly distributed

$$i \propto n_s (1 - n_s)^{1/2} \qquad (16)$$

gives the implicit dependence of n_s on the current[7]. In general n_s is a decreasing function of i up to the critical current where for a link driven by a current source n_s drops abruptly to zero and L_k becomes very large. (That is, the current is being carried by a few superelectrons moving very fast and thus having large kinetic energy.) Rather than $L_k(i)$, it is convenient to determine the phase shift $\Theta(i)$ across the link. Using Eq. (15) we can write Eq. (14) as

$$\frac{\Phi_o}{2\pi} \Theta(i) + iL + \phi_x = 0 \ . \qquad (17)$$

With a magnetometer loop encircling the ring we can measure the total flux $\phi = iL + \phi_x$ in the ring and calculate Θ from Eq. (17). Further, if ϕ_x is applied with a solenoid within the ring, it can be precisely known and if L is known, the current i can be determined too. Although this experiment has not yet been done very precisely, the general features of the results are shown in Fig. 5, where the singlevalued function $i(\Theta)$ has been plotted. As the link is weakened, that is if some mechanism is used to intrinsically reduce n_s in the link, $i(\Theta)$ takes the form of the lower curve and approaches $i = i_c \sin \Theta$ for a very weak link[8]. (The portion of the curve to the right of the peak is not accessible in an experiment that fixes the current, but can be determined in one that fixes the phase.)

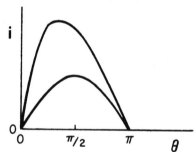

Fig. 5

Current-phase relation for weak links.

It is useful to look at the current i as a function of the applied flux ϕ_x. In Fig. 6 the current has been plotted for a ring with a Josephson junction (so that $i = i_c \sin \Theta$) for various values of L. From Eq. (17) it is clear that for sufficiently small L, $\Theta \sim 2\pi\phi_x/\Phi_o$ so that $i \sim \sin(2\pi\phi_x/\Phi_o)$ as shown by curve L_1. For larger L (curves L_2 and L_3) when i reaches i_c the n = 1 quantum state has lower energy than n = 0 and there is an abrupt jump in current corresponding to a jump to n = 1.

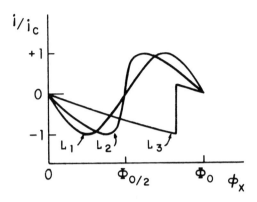

Fig. 6

Current in ring containing a Josephson junction as a function of externally applied flux.

This behavior provides the basis for one very successful magnetometer for which a superconducting ring containing a weak link is driven with an alternating flux ϕ_x. If the amplitude is just large enough to cause a jump (for example, slightly less than Φ_0 in Fig. 6 for curve L_3), the voltage e induced in a pickup coil around the ring will be a series of pulses as shown in Fig. 7.

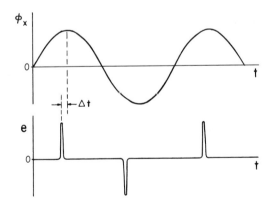

Fig. 7

Voltage in pickup coil around ring to which a sinsusoidally varying flux is applied.

The position of these pulses in time with respect to the $\phi_x(t)$ waveform (shown as Δt in Fig. 7) changes if a steady field is applied to the ring and the change is a measure of the applied field. In practice various electronic techniques are used to extract this information and flux changes less than $10^{-4}\ \phi_o$ can be detected. A detailed discussion of this type magnetometer will be given by Professor Webb.

To determine the behavior of a weak link as a circuit element we need to know the relationship between current through the link and the voltage across it. Returning to the ring experiment, consider the situation when ϕ_x is changing in time thus producing an emf in the ring. If L is very small we find by taking a time derivative of Eq. (17) that

$$V_{link} = \frac{\Phi_o}{2\pi} \dot{\Theta} . \qquad (18)$$

This relation together with the function $\Theta(i)$ determined from the static experiment completely determine the supercurrent behavior and can be considered a pair of parametric equations giving the relationship between the voltage and the current. For very weakly coupled superconductors these are equivalent to the Josephson equations [Eqs. (12) and (13)].

Many techniques are used for producing weakly linked superconductors. The most common ones are shown in Fig. 8. The usual way of characterizing a weak link is by measuring its dc voltage-current curve and for all weak links it has the general features shown in Fig. 9. Up to a critical current i_c, current flows with no voltage. At higher currents there is a voltage and at very large currents the slope of the V-I curve is just the normal state resistance. For tunnel junctions the V-I curve shows hysteresis and as the current is decreased the voltage follows the upper curve in Fig. 9 even for current less than i_c. For small SNS junctions there is no hysteresis and the lower curve is reversible. Other types of weak links show intermediate behavior as indicated in Fig. 9. Bridges often have V-I curves that do not extrapolate to zero current and do not break away so abruptly at i_c. The departures of the V-I curves from simple resistive behavior above i_c indicate that the supercurrent continues to play a role even at high current. In particular we can see from Eqs. (12) and (13) that when there is a voltage across a Josephson junction there is an alternating supercurrent

$$i = i_c \sin \frac{2e}{\hbar} Vt . \qquad (19)$$

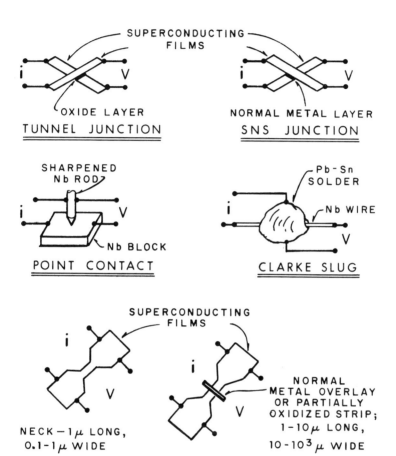

Fig. 8

Types of weakly linked superconductors.

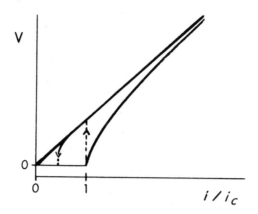

Fig. 9
Typical V-I curve for weakly linked superconductors.

with frequency

$$\nu = \frac{2e}{h} V \quad . \tag{20}$$

Thus even with a steady voltage or current applied to it, a weak link contains time varying currents and fields and it is necessary to consider the other conduction processes present in particular types of links[9]. Some lumped circuit representations of the various types of weak links are shown in Fig. 10. The symbol in Fig. 10(a) represents the purely supercurrent properties of the weak link which are described by the pair of equations

$$i = i(\theta) \quad ,$$

$$V = \frac{\Phi_o}{2\pi} \dot{\theta} \quad .$$

For all sufficiently weak links $i(\theta)$ is the Josephson relation

$$i = i_c \sin \theta \quad .$$

In Fig. 10(b) the resistor R accounts for the shunt conductance of the normal electrons. This circuit is a good representation of an

SNS junction. Bridges and point contacts are better represented by 10(c) which includes a series inductance which in these cases is appreciable with respect to the effective inductance of the junction. A tunnel junction with its large shunt capacitance is best represented by 10(d), where, in fact, the inductance may be negligible. However, a general weak link will include all the elements of Fig. 10(d), although the effects of L and C may be expressed as an effective L' or C' alone. Values for these equivalent circuit elements can be determined from the dc V-I curve from measurements of i_c, R and the current intercept of the extrapolated linear portion of the curve (or the current value at which the voltage jumps back to zero voltage when there is hysteresis.) With these parameters it is then possible to analyze the behavior of weak links in various kinds of circuits.

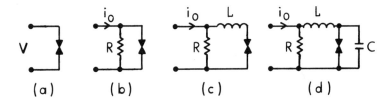

Fig. 10

Equivalent circuits for various types of weak links.

It is very informative to examine the instaneous behavior of the supercurrent for each of these models[10]. For the case of Fig. 10(a) with a voltage across a Josephson junction we have already seen that there is an oscillating supercurrent given by Eq. (19), but the average supercurrent is zero. If the circuit of Fig. 10(b) is driven with a constant current i_o,

$$(i_o - i)R = \frac{\Phi_o \dot{\theta}}{2\pi} = \frac{\Phi_o}{2\pi i_c} \frac{1}{\sqrt{1-(i/i_c)^2}} \frac{di}{dt} \quad . \tag{21}$$

Written in terms of the supercurrent i, Eq. (21) shows that the Josephson junction behaves like a current dependent inductor whose scale is determined by $\Phi_o/2\pi i_c$. Usually the equations for the models are expressed in terms of θ and the equation for the general case including L and C have the same form as the equation for a damped pendulum driven with a constant torque. This led to some useful mechanical analogs[10].

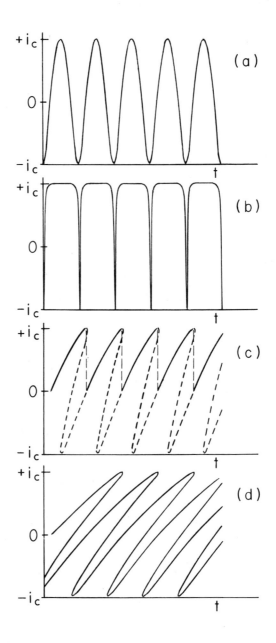

Fig. 11

Typical waveforms for the supercurrent in various kinds of weak links.

In Fig. 11 calculated values of the current in the link are plotted for various choices of the lumped circuit elements in Fig. 10 assuming that the circuits are driven with a constant current $i_o \gg i_c$. The supercurrent oscillates nearly sinusoidally [Fig. 11(a)] for a junction shunted by a small resistor [Fig. 10(b)]. For a larger value of resistance the oscillation is very peaked [Fig. 11(b)] and there is an average supercurrent through the link. A similar waveform results when a shunt capacitor is added, however as the capacitance is increased, the oscillation becomes more nearly sinusoidal because the capacitor keeps the voltage more nearly constant across the junction. Another feature also appears with the presence of the capacitor: Although there is no oscillation for $i_o < i_c$ as i_o is increased from zero, once the oscillation has been initiated by raising i_o above i_c, it persists for $i_o < i_c$ as i_o is reduced. It finally subsides when the current i_o can no longer supply the energy being dissipated each cycle. This feature accounts for the hysteresis in the V-I curve (Fig. 9) of tunnel junctions. In fact, the dc V-I curve is obtained by calculating the average voltage across the circuit as a function of i_o.

If instead of a capacitor an inductor is placed in the circuit [Fig. 10(c)], the oscillation has the form shown in Fig. 11(c). There are abrupt jumps in the current i corresponding to sudden changes in the phase difference across the junction by 2π. As the inductance L is increased, the waveform tips more. The oscillation in Fig. 11(d) represents a case where a sudden jump of 2π or 4π is possible. For large enough i_o, the jump will be to the state nearest zero current. This represents a relaxation-type oscillation and is the case observed for many point contacts and bridges. One way of picturing this type of oscillation is shown in Fig. 12. As the current increases, n_s is much reduced and suddenly one (or more) loops of the helix is pinched off, corresponding to a jump in phase of 2π (or some multiple of 2π). The current readjusts and the process repeats.

Waveforms of the types shown in Figs. 11(b), 11(c) and 11(d) can contain high frequency components corresponding to very high harmonics of the fundamental frequency or repetition rate ν_r.

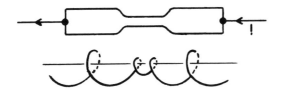

Fig. 12

Representation of the macroscopic wave function along a weak link.

Also for the relaxation-type oscillation the average voltage V across the weak link is

$$V = m\Phi_o \nu_r \tag{22}$$

where $m(2\pi)$ is the phase jump each cycle.

Although the equivalent circuits in Fig. 10 represent the intrinsic properties of a link, an actual lumped parameter circuit can be used to produce this behavior at lower frequencies. The results of an experiment of this type are shown in Fig. 13[10,11]. A tank circuit (LC) resonant at 60 MHz was connected across a point contact with a series inductance $L \sim 10^{-10}$H and shunt resistor $R \sim 10^{-5}\Omega$ and the output from the tank circuit plotted as a function of bias voltage V. The peak at $V/V_o = 1$ corresponds to a relaxation oscillation with $\nu_r = 60$ MHz; the peak at 1/2 corresponds to an oscillation with $\nu_r = 30$ MHz but with a large second harmonic content to which the tank circuit responds. The data shown are actually for an oscillation with $m \sim 30$ and peaks for $V/V_o \sim 1/50$ could be observed[10].

It appears then that the behavior of a superconducting weak link in many devices can be analyzed by using the V-I curve to

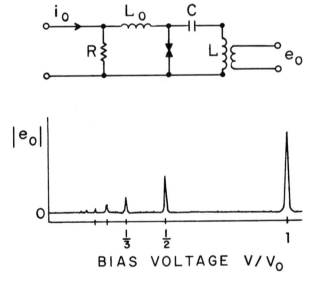

Fig. 13

Output from tank circuit coupled to a voltage biased point-contact.

determine the parameters in an equivalent circuit of the type shown in Fig. 10(d) and then using the circuit as a basis for calculations.

We can briefly examine several more important features of weakly linked superconductors. Consider what happens when in addition to a steady current i_o there is an additional current $i_1 \cos \omega t$. For small R and L = 0 this is essentially equivalent to applying a voltage $V = V_o + V_1 \cos \omega t$ to the junction. Then from Eqs. (12) and (13) we have

$$i = i_c \sin \left[\frac{2e}{\hbar} \left(V_o t + \frac{V_1}{\omega} \sin \omega t \right) + \Theta_o \right]$$

$$= i_c \sum_{k=-\infty}^{k=\infty} J_k \left(\frac{2eV_1}{\hbar \omega} \right) \sin \left[\left(\frac{2eV_o}{\hbar} + k\omega \right) t + \Theta_o \right] . \quad (23)$$

This result accounts for the character of the V-I curve of a weak link when it is irradiated with microwaves giving in general a series of evenly spaced constant-voltage steps as shown in Fig. 14.

These steps are apparent as the dc contributions in Eq. (23) which occur for

$$V_o = \frac{k\hbar\omega}{2e} \qquad (24)$$

and as you have already seen are the basis for the use of the Josephson Effect in establishing a voltage standard.

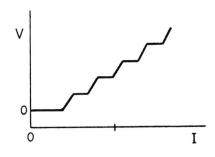

Fig. 14

Form of the V-I curve of a weak link irradiated with microwaves.

From Eq. (23) with k = 0 we see that the dc zero-voltage current (i.e. for $V_o = 0$)

$$i = i_c J_o\left(\frac{2eV_1}{\hbar\omega}\right) \sin \theta_o , \qquad (25)$$

is sensitive to the presence of voltages of all frequencies and thus provides the mechanism for a broad band radiation detector. Also the size of each of the dc current steps in Fig. 14 has a Bessel function dependence on the amplitude V_1 of the alternating voltage, hence these steps can be used for frequency selective detection. Furthermore, there are oscillating currents present in the weak link at the sum and difference frequencies, $2eV_o/\hbar \pm k\omega$ so the weak link can be used as a local oscillator and mixer.

In general then a biased weak link is both a non-linear inductor and an oscillator and has an amazing variety of applications many of which will be discussed by S. Shapiro who was first to observe the microwave induced steps and has pioneered in the study and application of these effects.

So far we have been discussing almost exclusively the time dependence of the supercurrent in weak links. Now let us consider very briefly the spatial distribution of the supercurrents.

In a sufficiently small ideal tunnel junction the current is uniformly distributed across the junction when there is no applied field (Fig. 15, H = 0). However, when there is an applied field perpendicular to the junction (into the paper in Fig. 15), since the phase difference across the junction depends on the magnetic field [see Eq. (3)] the current density varies across the junction. This variation is depicted in Fig. 15 for several values of the applied field. For $H = H_o$ the difference in phase between the two edges is exactly 2π, and one flux unit Φ_o is present within the junction; that is, $\phi_x = H_o (\lambda_1 + \lambda_2 + t)w = \Phi_o$ where t is the thickness of the barrier, w its width, and λ_1 and λ_2 the penetration depths in the two superconductors. For this case there is no net supercurrent through the junction. The curve in Fig. 15 shows the variation of the maximum zero-voltage current with applied field,

$$i_c = i_o \left|\frac{\sin(\pi \phi_x/\Phi_o)}{\pi \phi_x/\Phi_o}\right| . \qquad (26)$$

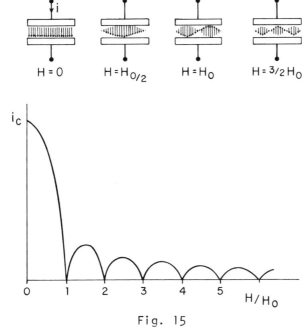

Fig. 15

Effect of a magnetic field on the current distribution and critical current of a tunnel junction.

In general the distribution of current in the junction can be calculated from the Josephson relation with a position dependent phase difference $\Theta(z)$,

$$j(z) = j_c \sin \Theta(z) \tag{27}$$

and using the fact that the gradient of the phase depends on the local field so that

$$\frac{\partial \Theta}{\partial z} = \frac{2\pi}{\Phi_o} d\, H(z) \tag{28}$$

where d is twice the penetration depth λ plus the insulator thickness and z is position along the barrier[12]. For a small junction the Fraunhofer pattern Eq. (26) for the dependence of the total current on the applied field results. For junctions large with respect to the Josephson penetration depth,

$$\lambda_J = \left(\frac{\hbar}{2\mu_o j_c ed} \right)^{1/2}, \tag{29}$$

the results are more complicated and depend on the fields produced by the currents. In general the current direction alternates along the junction and corresponds to a line of current vortices in the barrier.

This seems an appropriate point to mention another important type of magnetometer which uses two weak links connected in parallel (Fig. 16). The maximum zero-voltage current i_{max} through the circuit depends on the flux enclosed in the loop joining the links since, in addition to the contributions due to the two junctions, the phase difference depends on the $\oint \bar{A} \cdot d\bar{s}$ around the loop. In this case

$$i_{max} = i_o |\cos \pi \phi_x/\phi_o|, \tag{30}$$

where ϕ_x = HA and A is the area of the loop. The maximum current is thus periodic with period $H_o = \Phi_o/A$ as shown in Fig. 16(a). In general, if the weak links were identical Josephson junctions, there would be an additional dependence on the field for each junction and the interference effect [Eq. (30)] would be modulated by the diffraction pattern [Eq. (26)]. For use as a magnetometer the device is usually biased with some current $i_o > i_c$ and the periodic variation of the voltage used to sense field changes. Just as for the ring containing a single weak link, various electronic techniques can be used to detect field changes of about 10^{-10} Gauss.

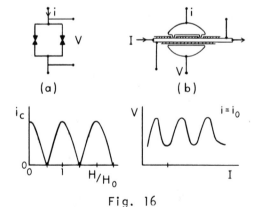

Fig. 16

Interference effect in parallel weak links.

The Clarke Slug is an interesting and useful variant of this device[13]. Shown in a cutaway view in Fig. 16(b) is an idealized Slug showing weak links near the edges of the solder blob. This device is essentially the device of Fig. 16(a) with the area between the links provided by the thickness of the insulation (shown cross hatched in the Figure) on the wire. Magnetic flux linking the area between the weak links is produced by a current I flowing along the wire; thus the voltage across the device for a fixed bias current i_o is a periodic function of I. This device is basically a superconducting current meter and can be converted to a magnetometer by making a superconducting loop of the wire. Likewise the devices that are basically magnetometers can be used as ammeters by placing a coil inside the magnetometer loop.

To return to some examples of the spatial distribution of the supercurrent, consider a thin superconducting film with a magnetic field applied perpendicular to the film. In this geometry even for type I materials the field enters the film as an array of isolated flux lines as shown in the lower part of Fig. 17. Associated with each line is a small essentially normal core ($n_s = 0$) surrounded by a current vortex (upper part of Fig. 17). Evaluating the fluxoid around any path encircling one of these cores gives n = 1.

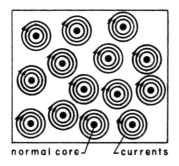

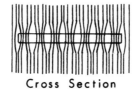

Cross Section

Fig. 17

Field and current distributions in thin superconducting film with applied magnetic field perpendicular to the surface.

If a current is now passed through the film, the vortices experience a force at right angles to the current direction. If the current is sufficiently large, the force will exceed the pinning force on the vortices and they will move through the film producing a voltage across it.

An interesting demonstration of this flux flow voltage is Giaever's dc transformer which uses a secondary film insulated from the first film by perhaps 1000 Å of oxide[14] (Fig. 18). The same voltage appearing across the primary is found across the secondary film and can be pictured as shown in Fig. 18 as discrete flux lines moving through both films.

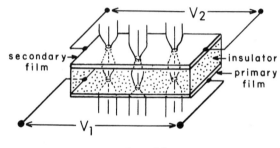

Fig. 18

Coupled flux flow in superimposed thin films.

This flux flow picture is also a valid way of thinking of an SNS junction and wide bridges. The vortices move along the barrier and the average voltage across the junction is given by $V = \Phi_o \nu$ where ν is the number of vortices crossing the weak link each second. For point-contacts and microbridges the picture is less useful since the dimensions become small with respect to a localized vortex; for these the notion of pinching off loops of the helix mentioned earlier seems more appropriate. In any case, the link can be regarded as a quantized flux gate. This property together with the fast switching times of weak links makes unique new kinds of computer elements a possibility using arrays of weak links for storage and logic.

There is an interesting synergism among some of these devices. It appears that full use of the extremely high sensitivity of superconducting magnetometers and multimeters may be possible only by using the near perfect shielding provided by superconductors for electromagnetic fields and the stability and low noise provided by the low temperature environment. Further, for magnetic measurements, only a truly persistent superconducting solenoid provides the field stability required to capitalize on the ability of superconducting magnetometers to detect small changes in field in the presence of large ambient fields.

Even without these ultimate measures, superconducting devices are opening new realms for experimental physics and technology.

REFERENCES

1. A survey of the literature from 1959 to March 1967 was compiled by W. S. Goree and E. A. Edelsack and issued as a cooperative effort of Stanford Research Institute and the Office of Naval Research. This survey has been continued through a joint publication of the National Bureau of Standards and the Office of Naval Research issued quarterly. It is entitled "Superconducting Devices and Materials" and is currently being compiled by W. S. Goree, E. E. Takken, R. A. Kamper and Neil A. Olien. Copies can be obtained from Cryogenic Data Center, National Bureau of Standards, Boulder, Colorado 80302.

2. F. London, Superfluids, Vol. I (Dover Publications Inc., New York, 1961) p. 82.

3. J. Bardeen, Phys. Rev. Letters $\underline{7}$, 162 (1961); J. E. Mercereau and L. T. Crane, Phys. Letters $\underline{7}$, 25 (1963).

4. W. L. Goodman, W. D. Willis, D. A. Vincent and B. S. Deaver, Jr. Phys. Rev. B (to be published).

5. W. A. Little, Proceedings of the Symposium on the Physics of Superconducting Devices, University of Virginia, 1967.

6. B. D. Josephson, Superconductivity, R. D. Parks, ed. (Marcel Dekker, New York, 1969); P. W. Anderson, Progress in Low Temperature Physics, Vol. V, C. J. Gorter, ed. (North-Holland, Amsterdam, 1967).

7. P. G. De Gennes, Superconductivity of Metals and Alloys, (Benjamin, New York, 1966) p. 184.

8. A. Baratoff, J. A. Blackburn, B. B. Schwartz, Phys. Rev. Letters $\underline{25}$, 1096 (1970); A. H. Silver and J. E. Zimmerman, Phys. Rev. $\underline{157}$, 317 (1967); W. D. Willis, Flux Distribution and Quantized Flux States in Superconducting Cylinders, M.A. thesis, University of Virginia, 1971.

9. W. C. Stewart, Appl. Phys. Letters $\underline{12}$, 277 (1968); D. E. McCumber, J. Appl. Phys. $\underline{39}$, 3113 (1968); J. R. Waldram, A. B. Pippard, and J. Clarke, Phil. Trans of the Roy. Soc.,

London **268**, 265 (1970); P. K. Hansma, J. N. Sweet and G. I. Rochlin, Phys. Rev. B (to be published).

10. D. B. Sullivan, R. L. Peterson, V. E. Kose, J. E. Zimmerman, J. Appl. Phys. **41**, 4865 (1970).

11. J. M. Pickler, Modes of Oscillation of a Point-Contact Coupled to a Resonant Circuit, M. A. thesis, University of Virginia, 1971.

12. C. S. Owen and D. J. Scalapino, Phys. Rev. **164**, 538 (1967).

13. J. Clarke, Phil. Mag. **13**, 115 (1966).

14. I. Giaever, Phys. Rev. Letters **15**, 825 (1965).

SUPERCONDUCTIVITY IN DC VOLTAGE METROLOGY

T. F. Finnegan

Institute for Basic Standards, National Bureau of Standards

Washington, D.C. 20234

INTRODUCTION

In recent years, the use of the ac Josephson effect to determine 2e/h has assumed an important role in increasing our knowledge of the fundamental physical constants. In particular, when combined with values of certain other physical constants, a determination of 2e/h yields an accurate value of the fine structure constant which can be used to test quantum electrodynamics.[1] The ac Josephson effect via 2e/h can also be used to maintain a unit of voltage, or more precisely electrochemical potential. It is this latter technological application which we consider in detail here. The future role of the cryogenic potentiometer (voltage divider) and superconducting voltmeter (galvanometer) in a Josephson-effect voltage standard is also explored, and a possible secondary voltage standard based on Giaver tunneling is described.

All Josephson effect measurements of 2e/h rely on the fact that if two pieces of superconductor (each characterized by a single quantum-mechanical phase) are weakly coupled in some fashion, for example, by the tunneling of quasiparticles through a very thin insulating barrier, and an electrochemical potential difference $\Delta\mu$ is maintained across the superconductors, the junction formed will carry an oscillating supercurrent with fundamental frequency

$$\nu_J = 2\Delta\mu/h. \tag{1}$$

If $\Delta\mu$ is identified with eV where V is the electrostatic potential difference (voltage), Eq. (1) becomes

$$\nu_J = 2eV/h. \tag{2}$$

This of course is the Josephson frequency-voltage relation and the effect is the ac Josephson effect. A measurement of the frequency-voltage ratio determines $2e/h$.

Mathematically, a macroscopic piece of superconductor can be represented by a complex quantum mechanical wavefunction. For a Josephson junction, the wave-functions of the two weakly-interacting superconductors overlap in such a way that a phase dependent supercurrent density j can flow across the junction. This supercurrent density can be written in the form

$$j = j_o f(\varphi) \tag{3}$$

where j_o depends on the strength of the coupling between the two superconductors, the temperature, the dc bias voltage, and the magnetic field; and $f(\varphi)$ is a periodic function of φ, the relative quantum mechanical phase difference across the junction. For a tunnel junction, a standard perturbation calculation yields

$$f(\varphi) = \sin \varphi, \tag{4}$$

and the time-dependent supercurrent density is just

$$j = j_o \sin \left(\frac{2\Delta\mu}{h} t + \varphi_o \right) \tag{5}$$

The origin of the radiation induced steps used in $2e/h$ measurements can be understood as follows: When exposed to radiation of frequency ν, an oscillating phase change can be introduced between the two weakly coupled superconductors forming the junction. In the simplest case, the supercurrent density (from Eq. (5)) is then given by

$$j = j_o \sin \left[\frac{2\Delta\mu}{h} t + \frac{2eV_{ac}}{h\nu} \sin(2\pi\nu t + \theta) + \varphi_o \right] \tag{6}$$

where V_{ac} is the maximum amplitude of the "induced ac voltage" across the junction. When $2\Delta\mu = nh\nu$ (n is an integer), the supercurrent density will have a time average or dc part

$$j_n = j_o (-1)^n J_n \left(\frac{2eV_{ac}}{h\nu} \right) \sin(\varphi_o - n\theta), \tag{7}$$

and the dc current-voltage (I-V) characteristic will exhibit a series of current steps at voltages $V_n = nh\nu/2e$. (Experimentally, the factor $\sin(\varphi_o - n\theta)$ can be varied by means of the externally applied dc current.) An example of these steps is shown in Fig. 1. At microwave frequencies, $\nu = 10$ GHz, the voltage spacing of adjacent steps is about 20 μV.

Before proceeding into a discussion of the relative merits of various Josephson junction devices for $2e/h$ measurements, it is

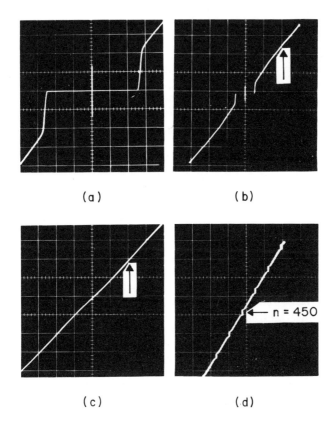

Figure 1. I-V characteristics of a Pb Josephson tunnel junction. (a) Vertical scale is about 2.5 mA/cm and horizontal scale is 1 mV/cm; (b) 10 mA/cm and 5 mV/cm; (c) also 10 mA/cm and 5 mV/cm; (d) 50 µA/cm and 25 µV/cm. The arrows in (b) and (c) indicate the voltage at 10 mV; the arrow in (d) indicates the induced step at about 10.2 mV corresponding to n = 450. The applied microwave frequency in (c) and (d) is approximately 11 GHz. (Reprinted from Reference 5).

useful to consider the question "What constitutes a good voltage standard?" The two most important requirements of a stable voltage standard are (a) a physical phenomena (or alternatively suitable artifact) exist which produces a reproducible potential difference independent of experimental parameters and (b) the technology must exist to compare the "standard voltage" to other sources of emf at the required level of accuracy.

The first of these two requirements is not particularly well met by the present voltage standard, the standard cell. The standard cell is an electrochemical system designed specifically for its properties as a stable source of emf. The particular type of cell in use today is the Weston or cadmium sulfate cell invented by Weston in 1892.[2] These cells have very large temperature coefficients and are sensitive to shock and vibration. The emfs of different cells are known to age or drift with time at different rates and are affected by the history of the individual cell. The usual procedure in maintaining a local voltage standard has been to use a group of cells and to intercompare them frequently. The two principle limitations in the use of cells has been (a) the inability to detect slow drifts in the mean of an entire group, and (b) the uncertainty which enters when as-maintained units of different laboratories are compared by shipping a group of cells from one site to another. The latter uncertainty (1σ) is typically one or two parts in 10^7. Standard cells can be routinely intercompared (over very short times) with a precision of about 7 parts in 10^9 so that the measurement technology itself is not a problem.

A Josephson-effect voltage standard, on the other hand, provides the basis for an "atomic" standard free of these difficulties. The Josephson junction can be considered a frequency-voltage transducer with a conversion ratio which is simply $2e/h$. In principle, any laboratory can set up the necessary apparatus and establish an operationally identical unit of voltage based on the Josephson effect. The Josephson relation is believed to be exact on theoretical grounds[3] to at least 1 part in 10^8 and a variety of experimental tests[4,5] of the relation have all verified its validity to better than 1 part in 10^7.

The major experimental problem in implementing a Josephson-effect voltage standard results from the relatively small usable voltages (V_n) which can be induced across a Josephson junction. In a single junction, these voltages are typically several millivolts and thus for calibration purposes, the junction output must be used in conjunction with a voltage divider to obtain about one volt, the emf of a conventional standard cell. On the other hand, the frequency measurement of the applied radiation can be readily made to 5 parts in 10^9 or better with straightforward and well established techniques.

THE JOSEPHSON DEVICE

Various methods have been developed to form a Josephson junction. The three types which have been used in precision 2e/h measurements are tunnel junctions,[5,6] point contacts,[7,8] and solder-blob junctions.[9,10] Josephson tunnel junctions are usually made by first evaporating (in a vacuum of about 10^{-6} torr) a thin film ($\sim$ 1500 Å thick) of superconducting metal (e.g., lead or tin) on a suitable substrate such as glass, then oxidizing the metal in an oxygen atmosphere either thermally or by means of a glow discharge to form a thin insulating barrier (about 10 Å thick), and finally evaporating a second film of superconductor. The area of overlap of the two films defines the junction.

The point contact junction is formed by pressing a superconducting point (usually Nb) against a second superconductor. The weak coupling is introduced through the extremely small area of contact at the point and can be varied in situ mechanically by adjusting the contact pressure. The relative ease of construction (no evaporation system is required) and the capability for continuous adjustment of the weak coupling during an experimental run are two very useful features of the device. This type of junction, however, is by construction, usually rather sensitive to external mechanical vibrations.

The slug or solder blob junction is made by surrounding an oxidized superconducting wire (usually Nb) with a small bead of solder. Two wires (usually copper) are imbedded in the solder and together with the two ends of the Nb wire form the four terminals of the device. It is the simplest of the three to construct.

The tunnel junction has proved to be the most successful of these three types for obtaining large step voltages V_n. For a 2e/h measurement, it is not just the voltage V_n which is important but the total height of the current step. A step height of approximately 10 μA is about the minimum acceptable for precise 2e/h measurements. (This minimum, however, is a strong function of (a) external noise which tends to round the edges of the step, (b) the stability of the incident microwaves whose amplitude enters via v_{ac} of Eq. 7, and (c) the stability of the dc bias supply. The latter two must be stable long enough to permit the necessary galvanometer balances as part of the voltage measurement.) With a <u>single</u> point contact or slug it is very difficult to achieve large amplitude induced steps above about 2mV. Furthermore, the operation of several (e.g., 2 or 3 point contacts in series to obtain a substantially larger voltage poses a serious practical problem if each contact is to be adjusted independently at liquid helium temperatures. It is also not practical to put several slugs in series since it is virtually impossible to match the properties of each device (particularly the microwave frequency response) sufficiently well. With tunnel junction devices, on the other hand, usable steps can be obtained at voltages greater than 10 mV with a <u>single</u> junction and it is practical to operate several in series. An important advantage of operating several in series is that, for some fixed <u>total</u> (Josephson) voltage V_J, the step heights of each of the several junctions in series can all be substantially greater than the height of a single step at V_J.

The geometry of a rectangular Josephson tunnel junction is sketched in Fig. 2. The tunneling current, I, ideally flows through the insulating barrier (of thickness ℓ). The junction area is $L \cdot W$. The length d is the characteristic length that an external magnetic field will penetrate into the region about the barrier. For identical superconductors,

$$d = \lambda \left[\coth (t_1/\lambda) + \coth (t_2/\lambda) \right] + \ell \qquad (8)$$

where λ is the London penetration depth (a measure of the distance that a magnetic field will penetrate into the bulk superconductor) and t_1 and t_2 are the respective film thicknesses of the two sides

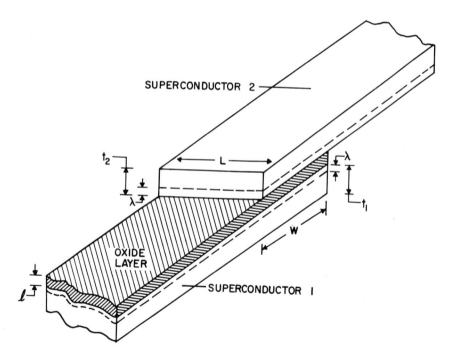

Figure 2. Ideal rectangular tunnel junction geometry.

of the junction. λ depends on the material [for Pb, $\lambda = 390$ Å; for Sn, $\lambda = 500$ Å] and is a strong function of temperature near T_c, the transition temperature of the film. For $T < 0.3\, T_c$, however, the temperature dependence of λ is weak, differing from λ ($T = 0$) by less than 1%. If t_1 and t_2 are much greater than λ, $d = 2\lambda + \ell$ as indicated in Fig. 2.

A Josephson tunnel junction can be regarded as an open circuited section of parallel plate transmission line and thus has resonant modes.[12] For the rectangular geometry indicated in Fig. 2, the fundamental resonances occur at frequencies

$$\nu_o(L) = \bar{c}/2L \tag{9a}$$

and

$$\nu_o(W) = \bar{c}/2W, \tag{9b}$$

where

$$\bar{c} = (\ell/\epsilon_r d)^{1/2} \tag{10}$$

is the phase velocity in the barrier between the two films of the junction. (ϵ_r is the relative permittivity of the barrier.) The phase velocity $\bar{c}$ is typically about 0.05 c and reflects the fact that the magnetic field associated with an electromagnetic wave can penetrate into the superconducting films whereas the electric field only penetrates into the insulating barrier.

The standard method for coupling microwaves into a tunnel junction is to mount the device in a section of waveguide. The characteristic impedance of the junction is usually very low compared to that of the waveguide, and thus there is a poor match between the junction and the microwave source except for frequencies near the resonant frequencies [$\nu_o(L)$ and $\nu_o(W)$] of the tunnel junction. The typical width of this resonant frequency response is a few percent corresponding to a "Q" of order 100.

Some of the problems involved in successfully achieving large induced-step voltages with either a single junction or several in series can be readily understood with the aid of Eq. 7. (For microwave induced steps in a resonant tunnel junctions, Eq. 7 is not valid since the spatial variation of the microwave fields in the barrier region and the dc voltage dependence of j_o have been neglected; however, qualitatively it does account for the observed behavior.) For a single junction, as the step number n is increased to achieve a larger V_n, the corresponding maximum step height decreases while the applied power must be continuously increased to achieve successive-step maxima.

We now may ask, "What is the maximum dc voltage at which usable steps can be induced in a single junction?" Usable steps at 10 mV with heights of 20 μA (in Pb-Pb oxide-Pb junctions) have already been used in precise 2e/h measurements. A partial answer to this question may lie in the experiments of McDonald et al[12] on high order induced steps in a Nb-Nb point contact. Their results indicate that for dc bias voltages above the phonon modes (particularly the longitudinal mode), the step amplitudes are strongly attenuated. For Nb, the peaks in the density of states for the transverse and longitudinal modes appear at dc bias voltages of 8 and 12 mV, respectively. For Pb, the phonon modes are at 6.5 mV and 11 mV. A usable device voltage of between 11 and 12 mV may therefore actually be the upper limit attainable with a single (Pb-Pb oxide-Pb) tunnel junction.

The use of several junctions in series involves different difficulties. The principle one is the requirement that the resonant dimension (i.e., the length L) of each be matched to about 1%. At X-band (ν = 10 GHz), L $\simeq$ 0.8 for a Pb-Pb oxide-Pb junction and a typical area (L·W) would be about 2 x 10^{-3} cm^2. Another difficulty is the complex microwave response caused by the direct coupling between junctions. Experimentally, up to 3 junctions have been simultaneously operated in series to yield a total voltage of 10 mV.[5]

The barriers for all of these junctions were formed by thermal oxidation of the first deposited film. The major difficulty with such devices is that they deteriorate rapidly (within a day or two) at room temperature (presumably due to diffusion of the metal through the thin oxide layer.) This problem has been partially solved by storing the devices at liquid nitrogen temperatures when not in use and by applying a layer of photoresist to protect them from the atmosphere during cycling through room temperature.[5] Recent work on a simple method of fabricating Nb-Nb oxide-Pb tunnel junctions has been reported, however, in which junctions were made by glow discharge oxidation of a sputtered Nb film followed by deposition of the top Pb film, and were then successfully stored at room temperature for more than 9 months.[13] The capability of fabricating Josephson junctions which are both useful for 2e/h measurements and which can be readily stored at room temperature for long periods of time will be an important step in making a Josephson-effect voltage standard more readily available to laboratories throughout the world.

In concluding this discussion on resonant tunnel junctions, it should be emphasized that no attempt has been made to be complete in this brief treatment of Josephson tunneling. Other parameters such as the Josephson penetration depth (λ_J), the temperature and magnetic field dependence of j_o, and self-magnetic field effects also play a

SUPERCONDUCTIVITY IN DC VOLTAGE METROLOGY

role in determining the detailed microwave response of a given tunnel junction. None of these parameters, however, affect the exactness of the frequency-voltage relation. Usable voltages of 10 mV are achievable with tunnel junction devices and have played a crucial role in the establishment of a Josephson-effect voltage standard with an accuracy of 3 parts in 10^8 (1σ) and an inherent precision of about 1 part in 10^8.

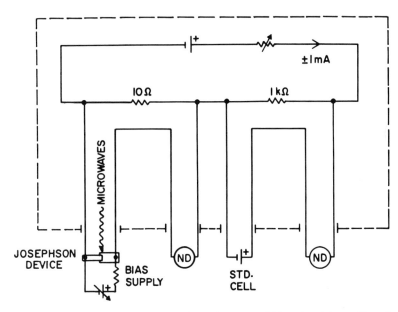

Figure 3. Simplified circuit diagram illustrating fixed voltage ratio technique used in measuring 2e/h.

LOW VOLTAGE COMPARATOR TECHNIQUES

In the early precision 2e/h measurements, the junction voltage V_J was treated as a conventional unknown emf of order 1 mV. These measurements[5,8] accurate to a few parts in 10^6, were limited primarily by uncertainties in the modified commercial potentiometers used to determine V_J. A drastic improvement in both the accuracy and precision with which 2e/h could be determined was made possible by an order of magnitude increase in V_J to 10 mV and by construction of special purpose instruments designed specifically for this application. Two independent techniques have been developed to obtain a fixed 100:1 voltage ratio between 10 mV and 1 V; and two instruments based on these techniques have been shown by direct comparison experiments[5] to agree within about 1 part in 10^8. These two methods will be described and their potential application to a cryogenic system will be examined.

The use of a fixed voltage ratio in 2e/h measurements depends on the fact that the Josephson device voltage is precisely tunable via the frequency of the incident microwave radiation. Consequently, the voltage comparator need have no adjustable resistance elements equivalent to the slidewire or dials of a conventional potentiometer with their attendant complexity and sources of uncertainty. The method of operation of a fixed ratio instrument is illustrated in Fig. 3. The working current (nominally 1 mA) through the two fixed resistors in series is adjusted so that the voltage across the larger resistance (1 kΩ) is equal to that of the standard cell. The Josephson device voltage (only one junction is indicated in Fig. 3 for conceptual simplicity) is then adjusted via the total step number N and the frequency ν so that it equals the voltage appearing across the smaller resistance (10 Ω). When both galvanometers have been nulled, the ratio of the two voltages is equal to the ratio of the two resistances.

This can be expressed mathematically as

$$V_s = \beta \, V_J \tag{11}$$

where V_S is the standard cell emf and β is the voltage (resistance) ratio with a nominal value here of 100. We can now define a quantity F_S, the standard cell Josephson frequency, as $F_S = \beta N \nu$ and rewrite Eq. (11) in the form

$$F_s = \frac{2e}{h} V_s. \tag{12}$$

F_S is the product of all the directly measured experimental quantities except the standard cell emf. Thus a given standard cell emf may be specified in terms of a frequency independent of any particular as-maintained unit of emf. An as-maintained unit using the Josephson effect can be readily implemented simply by arbitrarily selecting a best value of 2e/h (expressed in the old unit) at some point in time and thereafter attributing any observed changes in V_S to the standard cell.

Series-Parallel Method

One method for achieving the accurate resistance ratio (apparently first noted by Lord Rayleigh in 1882[14]) is to connect the same nominally-equal resistors alternately in series and in parallel. The simple theory upon which the series-parallel method depends is straightforward. If n nominally equal resistors are connected first in series and then in parallel, and Δ_i is the fractional deviation of the ith resistor from a nominal value, then the ratio of the series resistance R_s to the parallel value R_p is

$$\frac{R_s}{R_p} = n^2 [1 + \sigma^2 (\Delta_i)] \qquad (13)$$

where $\sigma^2 (\Delta_i)$ is the variance of the Δ_i. If the resistors are matched, for instance, to 1 part in 10^4, the ratio n^2 is accurate to about 1 part in 10^8. Of course, this simple analysis assumes ideal series and parallel connections. In general the resistance of these connections cannot be ignored.

The idea of establishing a resistance ratio using two series-parallel resistor networks appears to have been first used in a 10:1 resistance ratio device described by Wenner.[15] In that device two $n = 3$ resistor chains were connected to an intermediate resistor whose value was equal to the parallel resistance of either chain. (The intermediate resistor was necessary to establish a 10:1 rather than a 9:1 ratio.) By alternately exchanging the series and parallel connections of the resistor chains (making the necessary resistor connections and averaging the additional ratio measurements), the error introduced by the differences in the mean resistance of the two chains (ignoring lead corrections) was made negligible.

To establish a 100:1 voltage ratio using two resistor chains, the total number of (equal) resistors required per chain is $n = 10$. The choice of resistance value for these resistors depends on a number of factors. For large valued resistors the two principle limiting factors are the Nyquist noise voltage in the paralleled resistor string and the leakage resistance shunting the string in series. For small resistances, the limiting factors are self-heating (the absolute power dissipated in the resistors is inversely proportional to the resistance for a constant voltage) and lead and contact resistances of the connections. The choice of 100 Ω resistors represents a good compromise. The rms thermal noise voltage in the paralleled resistor string ($R/n = 10$ Ω) for a bandwidth of about 0.3 Hz (corresponding to a 1 sec. rise time) at 30 °C is approximately 0.2 nV (2 parts in 10^8 of 10 mV). Leakage resistances of greater than 10^{11} Ω are not difficult to maintain by careful design and the proper choice of insulators. Concerning self-heating the power dissipated in 1 kΩ series resistor is 100 times that dissipated in the 10 Ω paralleled resistor; however, the larger power, that dissipated in the 1 kΩ resistor chain with a current of 1 mA, is only 1 mW total (100 μW per resistor) and consequently the effects of self-heating in the resistors should be small. The use of 100 Ω resistors also makes the resistances of the paralleling connections less critical.

With 100 Ω resistors, elaborate paralleling connections are necessary to attain ratio accuracies of 1 part in 10^8 or better. The first step in constructing a series-parallel resistor network such as that of Hamon[16] is to connect the n resistors constituting a chain permanently in series as indicated in Fig. 4(a) by means of

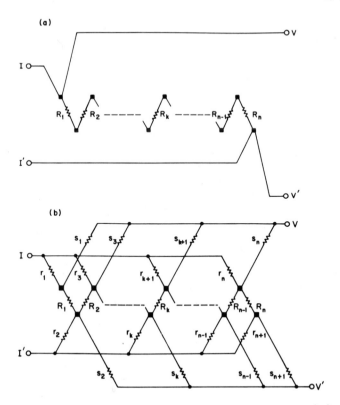

Figure 4. Generalized series-parallel network (a) Series connection; (b) parallel connection.

specially designed four-terminal (tetrahedral) junctions. These four-terminal junctions must be treated as four-terminal networks and constructed so that both the direct and cross resistances of the equivalent circuit are effectively zero.

The parallel connections of the n resistors are accomplished by adding compensating resistors (fans)[16,17] as indicated in Fig. 4(b). The purpose of the fans is to divide the current equally among the n resistors so that the voltages across them are equal. Figure 4(b) shows both current and voltage fans. The resistance of all the current fans except the two end fans is ideally r; that of the end fans, 2r. Similarly, for the voltage fans, the resistances are s and 2s. It is clear from the reciprocity theorem that if either set of fans are matched (assuming equal main resistors and perfect tetrahedral junctions), the parallel resistance will be exactly R/n.

The reproducibility of the contact resistance of (commercially available) rotary switches is about 1 mΩ. As a result, if they are used to make the paralleling connections of the resistor chains, the magnitude and degree of matching of the fans (which are in series with the switch contacts) should be such that the uncertainty associated with these contact resistance variations is negligible. The use of 20 Ω current fans (matched to 1 mΩ) and 10 mΩ voltage fans (also matched to 1 mΩ) can be shown (with the aid of Eq. 9 of the Ref. 18) to yield an error or order 10^{-8}. (With mercury-amalgam switches, a switch contact reproducibility of better than 10^{-6} is attainable; however, the large thermal emfs generated at these contacts are very undesirable and at low voltage levels require operation in an oil bath.)

A simplified circuit diagram of the double-series parallel exchange comparator built at the University of Pennsylvania is shown in Fig. 5. The largest source of uncertainty (about 2 parts in 10^8) was associated with the heating effects in the 100 Ω resistors caused by the difference in the power dissipation which occurs when the chains are switched from series to parallel.

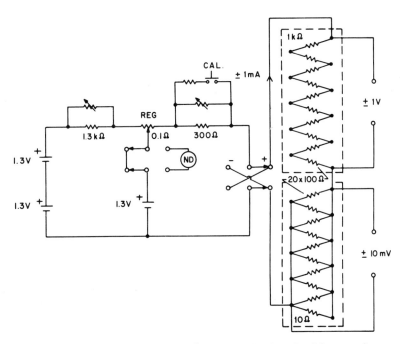

Figure 5. Simplified circuit diagram of the double series-parallel exchange comparator.

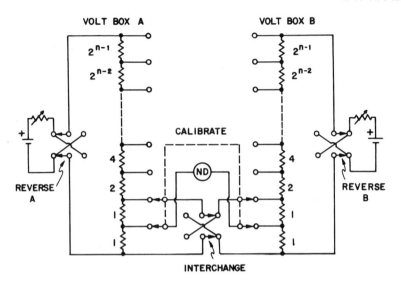

Figure 6. Circuit diagram illustrating generalized cascaded interchange calibration technique.

Cascaded-Interchange Method

The limiting factor associated with essentially all conventional voltboxes including current ones has been resistor heating, not thermal noise.[18] Thus, there has been no real need to optimize the subdivision of the voltbox for calibration purposes. The cascaded interchange method,[19,5] however, employs a voltbox which is optimized for self-calibration since thermal noise is ultimately the limiting factor in low voltage measurements at room temperature.

In the simplest case, the voltbox consists of a binary string of $n + 1$ resistors whose values are in the sequence 1, 1, 2, 4, ..., 2^{n-1} where $n \geq 1$ and the resistance ratio $2^n:1$ is the ratio of the total voltbox resistance to that of one of the two smallest valued resistors. (If $n = 8$, for example, the maximum resistance ratio and hence voltage ratio available is 128:1.) Each resistor (except the lowest valued one) has a trimmer to allow fine adjustment of its value.

Two similar, independently power voltboxes are used in the calibration procedure as illustrated in Fig. 6. The first step of this procedure is to form an equal-arm Wheatstone bridge out of the four lowest nominally equal resistors. If the resistors are adjusted to achieve a perfect null balance both before and after an interchange of the two arms in one voltbox, the pair of resistors in each voltbox will then have an exact unity ratio. Next, a new bridge is set up with the adjacent resistor (whose resistance is nominally equal to twice that of the lowest one) and the series connection of the two lowest value (matched) resistors to form a second nominal 1:1 ratio. This ratio is then made exactly equal to unity by adjusting the larger resistor in each string. Continuing in this manner, each resistor is matched to all those below it until the largest resistor is matched to the sum of all the others. Any integer ratio other than that of the strict binary sequence (for example, a 100:1 ratio) may be obtained by slightly modifying the resistor string. When comparing voltages, only one voltbox is required, but the second one can be used to check the results obtained with the first, since both are calibrated equally well.

One of the most important advantages of this technique is that the resistors can <u>always</u> be subjected to very nearly the same power during calibration, measurement, and even when not in use. The problem of self-heating during the voltbox calibration is thus avoided. The effects of differences in the resistances of the leads connecting the top and bottom of each of the calibration "bridges" are greatly reduced (to the extent that the voltages across the two voltboxes are equalized.) The two voltages can also be compared simultaneously minimizing the demands on the power supply. Finally, the use of a binary sequence of resistors results in the lowest calibration uncertainty for a given overall ratio in the ideal case where only thermal noise is considered. In the particular instrument based on this technique and used in 2e/h measurements[5], the limiting sources of uncertainty were (a) the random uncertainty associated with measurement and calibration and (b) switch and power supply variations during calibration.

The original aim in designing and building these two instruments was to determine 2e/h in terms of the NBS as-maintained unit of voltage and to demonstrate in practice the feasibility of maintaining a unit of voltage using the ac Josephson effect. The major limitation in achieving these goals was the standard cell itself. The final uncertainty (1σ) in the determination of 2e/h at the University of Pennsylvania was about 1 part in 10^7 and the demonstrated precision with which a group of standard cells could be maintained was about 4 parts in 10^8. The dominant uncertainty in determining 2e/h in NBS units was just the uncertainty in relating the local volt to the NBS volt.

During the summer (1971) a series of measurements have been carried out at NBS (with essentially the same apparatus as used in the Pennsylvania experiments) to determine 2e/h. The final result[20] is 2e/h = 483.593589 ± 0.000024 THz/V_{NBS-69}. This result together with the most recent results of the 2e/h groups at the National Physical Laboratory (Great Britain),[10] at the National Standards

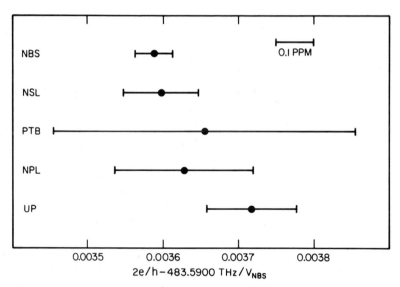

Figure 7. Comparison of 2e/h determinations (1971). All values have been expressed in NBS electrical units. The error bars are those quoted by the workers for their values of 2e/h expressed in their respective as-maintained national units.

Laboratory (Australia)[7] and at the Physikalisch-Technische Bundesanstalt (West Germany)[8] as well as the final University of Pennsylvania result are plotted in Fig. 7. All the measurements have been expressed in NBS units using the preliminary results of a series of direct standard cell intercomparisons which NBS conducted between BIPM, NPL, NSL, and PTB during the summer of 1971. The 2e/h results are in excellent agreement. The NBS, NPL, and NSL results (all made during the summer coincident with the standard cell transfers) agree to better than 0.1 ppm. These results also resolve an apparent discrepency of about 1/2 ppm in the relations of these national as-maintained units as determined via the 1970 triennial intercomparisons at the Bureau International des Poids et Mesures (see ref. 20).

CRYOGENIC VOLTAGE COMPARISON SYSTEMS

Perhaps the most obvious advantage of operating a voltage divider at liquid helium temperatures is the large reduction in thermal noise. A temperature of 1.3 °K, well below the superfluid transition point of liquid helium at 2.2 °K, can be readily achieved using standard techniques. At these temperatures the Nyquist noise in a 10 Ω resistor (Δf = 0.3 Hz) is only about 0.01 nV. Another important advantage is the capability of using superconducting wiring and switch contacts to reduce or eliminate sources of uncertainty associated with the finite resistance connections at room temperature. For example, the fans and tetrahedral junctions [Fig. 4(b)] necessary in the series to parallel transfer can be eliminated by using superconducting connections. Thermoelectric emfs nearly vanish in the low voltage (10 mV) portion of the circuit if a superconducting null detector is used since the entire low voltage circuit can be in a common helium bath, and the effects of the thermals in the 1 V circuit (which must emerge from the dewar for comparison with an external voltage standard) are reduced by a factor of 100. Leakage resistances should be less of a problem since conventional insulators usually improve at lower temperatures and superfluid helium itself is essentially a perfect insulator.

Certain problems present at room temperature remain, however, such as self-heating in the resistors. For although helium is a superfluid, resistors immersed in it are not thermally coupled substantially better than those in a room temperature oil bath due to the Kapitza boundary resistance[21] at the interface between a solid and superfluid helium. A number of new problems must also be considered. These include (1) the ohmic characteristics of a superconducting normal metal interface; (2) the development of stable precision resistors with small temperature coefficients ($\sim 10^6/°K$) at liquid helium temperatures (the materials commonly used to manufacture precision resistors for room temperature use such as Advance and Evanohm have large temperature coefficients of order $10^{-3}/°K$ at liquid helium temperatures); (3) the useful operation of superconducting null detectors with source impedances in the 1 to 10 Ω range (at present superconducting null detectors have a useful noise figure, i.e., a resolution limited by the Nyquisit noise in the source impedance, with a maximum impedance of order 10^{-1} Ω);[22] and (4) compatibility of the cryogenic voltage comparator system with the voltage sources to be measured. This last problem is particularly difficult if two Josephson devices (one providing the low voltage output, the other acting as detector[23]) are used in the same cryostat and each is irradiated at a different frequency. (The microwave power required for a 10 mV junction output is very large, typically about 1/2 W and consequently must be taken into account in constructing an all-cryogenic system since it directly affects the first 3 factors.)

On the other hand, a cryogenic version of the series-parallel comparator has several particularly attractive features. One is elimination of the elaborate paralleling connections required at room temperature. A second feature is the weak dependence of the series-parallel ratio (Eq. 13) on resistor matching. This is an important consideration if the resistors are to be matched at room temperature and successfully cycled to liquid helium temperatures. It is important to realize, however, that the mathematical identity for the ratio requires that the resistors be stable to first order over the period of an interchange. A promising resistance material composed of copper, silicon, and zinc with a temperature coefficient of less than 2×10^{-6}/°K at 2 °K has been reported.[23] The principle source of uncertainty in a cryogenic series-parallel divider appears likely to be the same as at room temperatures, i.e., self-heating of the resistors.

A cryogenic version of the cascaded interchange comparator also has several distinct advantages. The two limiting uncertainties at room temperature are reduced by more than an order of magnitude. (The one associated with lead resistance variations may be effectively eliminated.) The principle drawback in constructing such an instrument appears to be the practical problems associated with providing independent room temperature adjustment of all the main resistors in liquid helium.

Either of these methods applied at cryogenic temperatures has the potential to exceed the accuracy of the present (room temperature) instruments by a factor of 10. Many technical problems have to be investigated and solved, however, before a Josephson-effect voltage standard incorporating a cryogenic voltage comparison system can be successfully operated at the level of 2 parts in 10^8.

A TRANSFER STANDARD BASED ON GIAVER TUNNELING

Although 2e/h and the ac Josephson effect appears destined to become the primary maintenance voltage standard at NBS and other major national laboratories, the diverse equipment and time-consuming experimental procedures required for routine operation do not at present make it particularly useful as a transfer standard. Improvement in the coupling of microwave radiation into the junction devices and the development of cryogenic voltage dividers together with compatible superconducting galvanometers will make a portable Josephson effect voltage standard much more feasible. However, it is doubtful whether the expense of such a system will ever make it commercially competitive with the air enclosures of standard cells used at present. It appears possible that a shippable voltage standard (which might be used in a manner similar to conventional air enclosures of standard cells) may be constructed by taking advantage of the unique properties of superconducting tunnel junctions (Giaver tunneling) and sophisticated junction-fabrication techniques.[25]

The sharp increase in the quasi-particle tunneling current of a tunnel junction which occurs at voltages

$$V_o = \pm 2\Delta/e \qquad (14)$$

is the useful feature. Δ is the energy gap parameter of the superconductor (the two superconducting films have been assumed identical) and depends on both temperature and magnetic field. The magnetic field dependence presents no particular problem since the device can be shielded sufficiently well with either low permalloy shields at room temperature, superconducting shields at liquid helium temperature, or a combination of both. The temperature dependence is a more serious problem. The theoretical BCS temperature dependence of the energy gap parameter is given to a very good approximation[26] by

$$\Delta(T)/\Delta(0) = \tanh \{(T_c/T) [\Delta(T)/\Delta(0)]\} \qquad (15)$$

An order of magnitude calculation for Pb (T_c = 7.2 °K) near T = 1.3 °K yields a temperature coefficient for $\Delta(T)$ of about 3 parts in 10^7 per millidegree. There is also experimental evidence that the temperature dependence of $\Delta(T)$ in Pb is significantly weaker than this below about 0.7 T_c. With an electronic temperature controller operating in a liquid He bath below the λ point (and no appreciable heat load), it is a relatively straightforward procedure to regulate the temperature to a tenth of a millidegree.

The energy gap (2Δ) also has a finite slope primarily due to the finite quasi-particle recombination lifetimes, and gap anistropy. The magnitude of the dynamic resistance at the gap for a Pb-Pb oxide-Pb junction with an I_0 of about 2 mA is of order 50 mΩ. Thus, with a dc bias current of 1 mA stable and reproducible to 1 part in 10^6, a <u>nominal</u> voltage of 2.7 mV can in principle be established precise to about 2 parts in 10^8. The stable dc bias current may be attained with a power supply very similar to that used in the 2e/h voltage comparators. (This current can be easily and precisely reproduced to 1 part in 10^6 with a stable resistor and a standard cell.) For Pb, $2\Delta(0)$ = 2.7 mV so that a voltage of about 1 V must be obtained by operating a large array of junctions (about 380) in series. Devices with 462 junctions (each having an area of 0.5 x 0.3 mm) have been fabricated on a single (2.5 x 2.5 cm) substrate. Precise preliminary data on 3 junctions operated in series indicated that at the parts in 10^6 level (the overall resolution of the measurements) a voltage can be maintained in the manner discussed.

A voltage standard based on Giaver tunneling must be used as a secondary standard in a manner similar to conventional groups of standard cells; i.e., the nominal emf of the superconducting reference standard must be calibrated in terms of another standard. Thus, a Josephson-effect voltage standard would be required in conjunction with this secondary standard. It is the potential low cost and simple operation of this type of standard[25] which could make it a possible alternative to the standard cell as a widespread transfer standard.

In conclusion, superconductivity appears destined to play an increasingly important role in the establishment, maintenance, and precise measurement of dc voltages.

References

1. B. N. Taylor, W. H. Parker, and D. N. Langenberg, Rev. Mod. Phys. 41, 375 (1969).

2. W. J. Hamer, NBS Monogr. 84 (1965).

3. D. J. Scalapino, in Proc. International Conference on Precision Measurement and Fundamental Constants, edited by D. N. Langenberg and B. N. Taylor (NBS Spec. Publ. 343, U.S. Government Printing Office, Washington, D. C., 1971).

4. J. Clarke, Phys. Rev. Letters 21, 1566 (1968).

5. T. F. Finnegan, A. Denenstein, and D. N. Langenberg, Phys. Rev. Letters 24, 738 (1970); Phys. Rev. B 4, 1487 (1971).

6. W. H. Parker, D. N. Langenberg, A. Denenstein, and B. N. Taylor, Phys. Rev. 177, 639 (1969).

7. I. K. Harvey, J. C. Macfarlane, and R. B. Frenkel, Phys. Rev. Letters 25, 853 (1970), and private communication.

8. V. Kose, F. Melchert, H. Fack, and H.-J. Schrader, PTB-Mitt. 81, 8 (1971).

9. B. W. Petley and K. Morris, Phys. Letters 29A, 289 (1969); Metrologia 6, 46 (1970).

10. B. W. Petley and J. C. Gallop, in reference 3; and private communication.

11. J. C. Swihart, J. Appl. Phys. 32, 461 (1961).

12. D. G. McDonald, K. M. Evenson, J. S. Wells, and J. D. Cupp, J. Appl. Phys. 42, 179 (1971).

13. K. Schwidtal, Bull. Am. Phys. Soc. 16. 400 (1971).

14. Lord Rayleigh, Phil. Trans. Roy. Soc. 173, 697 (1882).

15. F. Wenner, J. Res. NBS 25, 253 (1940).

16. B. V. Hamon, J. Sci. Instr. 31, 450 (1954).

17. C. H. Page, J. Res. NBS 69C, 181 (1965).

18. R. F. Dziuba and T. M. Souders, 1966 IEEE Int. Conv. Rec., Pt. 10, 17 (1966).

19. A. Denenstein, Thesis, University of Pennsylvania (1969).

20. T. F. Finnegan, T. J. Witt, B. F. Field, and J. Toots, in <u>Proc. of the Fourth International Conference on Atomic Masses and Fundamental Constants</u>, Teddington (1971), to be published.

21. P. L. Kapitza, J. Phys. USSR $\underline{4}$, 206 (1941).

22. J. Clarke, Phys. Today, p. 30 (August 1971).

23. J. E. Zimmerman, Paul Thiene, and J. T. Harding, J. Appl. Phys. $\underline{41}$, 572 (1970).

24. D. B. Sullivan, Rev. Sci. Instr. $\underline{42}$, 612 (1971).

25. T. F. Finnegan and A. Denenstein, Metrologia $\underline{7}$, 167 (1971).

26. G. Rickayzen, Theory of Superconductivity (Interscience, New York, 1965), p. 178.

ELECTRIC AND MAGNETIC SHIELDING WITH SUPERCONDUCTORS[*]

Blas Cabrera[†]

Department of Physics
Stanford University
Stanford, California 94305

W. O. Hamilton

Department of Physics & Astronomy
Louisiana State University
Baton Rouge, Louisiana 70803

I. INTRODUCTION

For many years superconductivity was felt to be a phenomenon which would always be merely a laboratory curiosity, devoid of any practical applications. Yet today we realize so many potential applications that we can hold a summer course on the science and technology of superconductivity and treat many of the applications as commonplace, their having already attained the status of review by the New York Times or Time magazine. We would like to discuss one of the lesser known applications which has not yet attained the Times science page but which has the potential impact on the making of sensitive measurements that the high field superconductors have had on the creation of large stable magnetic fields.

In fact by using superconducting shielding it is possible to consider making measurements in a region containing zero electric and magnetic fields at frequencies from d.c. up to 10^{10} Hz. We shall briefly review the properties of superconductors which enable them to be such perfect shields. We will then discuss measurement techniques in low field regions and our most recent work on magnetic shielding. Finally we shall mention some applications of these shielding techniques. The experiments would be extremely difficult or even impossible without the shielding.

II. THE PROPERTIES OF SUPERCONDUCTORS

The macroscopic properties of superconductors which bear on the

---

[*] Research supported by Air Force Office of Scientific Research.
[†] NSF Predoctoral Fellow

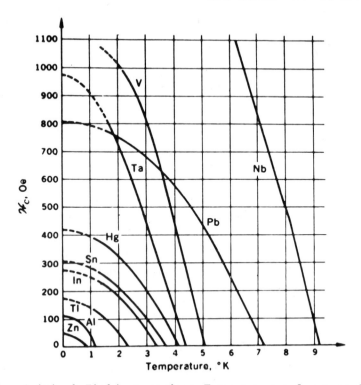

Figure 1. Critical Field at various Temperatures for some Superconducting Elements.

shielding problem are primarily the appearance of the zero resistance state and to a much lesser degree the Meissner effect.(10) The quantum effects have no bearing on the shielding against alternating fields but are crucial to the attainment of zero field.

The behavior of some superconducting metals is shown in Figure 1. This plot of the critical field vs. temperature is generally obtained by measuring the resistance of a superconductor with a very weak current while a magnetic field is applied perpendicular to the superconducting surface. The resistance is zero until the critical field is reached when a very sharp transition to a resistive state is observed. Here then we observe a first limitation on the use of superconductors for shielding. A superconductor behaves no differently than any other metal at fields above its critical field.

The existence of the critical field limits the size of currents which can be carried by the superconductor. When the current density is increased to the point that the magnetic field created by the current equals the critical field, the material reverts to the normal state. Thus an important consideration is the magnetic field at the surface of the superconductor due to both the external magnetic field and any currents induced in the shield. This magnetic field must be less than the critical field to keep the shield superconducting.

Merely describing a superconductor as a material with infinite conductivity is not sufficient to completely describe the superconducting state. London (11) gives a description which points out clearly why a superconductor is more than an infinite conductivity material. From Maxwell's equations:

$$\vec{\nabla} \times \vec{E} = -\frac{1}{c}\frac{\partial \vec{B}}{\partial t} \quad ,$$

and

$$\vec{E} = \frac{\vec{J}}{\sigma} \quad \text{(Ohm's law)} \quad .$$

Since we measure σ to be infinite then $\vec{E}$ must be zero in the superconductor. Thus Maxwell's equation tells us that B must be constant in time. If a changing external magnetic field is incident on a superconductor the superconductor should develop screening currents which create magnetic fields to oppose the change in B. This is simply Lenz's law. Since the material has no resistance the screening currents do not die out and the magnetic field in the metal remains constant.

The Meissner effect however adds more to this description. Meissner discovered that the magnetic field close to the outside of a superconducting body changed when the body became superconducting. He found that the superconductor tended to exclude the magnetic field which was passing through the material before it became superconducting. This cannot be explained by assuming perfect conductivity since we would expect the perfect conductor to simply trap the penetrating flux as it made the transition to the superconducting state. Thus Meissner showed that a superconductor also behaves like a perfect diamagnet, tending to exclude all magnetic fields from its interior.

It is experimentally very difficult to achieve a perfect Meissner effect. Most materials will exclude some of the magnetic flux but trap the rest in non-superconducting inclusions. The perfect conductivity of the material then makes it difficult to eliminate the trapped fields.

We now know, by the work of Deaver and Fairbank (4) and Doll and Naubauer (7), that the trapped magnetic flux exists in quanta. The flux quantum was measured to be:

$$\Phi_o = 2 \times 10^{-7} \text{ gauss cm}^2$$

and is the result of macroscopic long range order in the electrons of the superconductor. For any closed path entirely in the superconductor the magnetic flux linked by the path must be an integer multiple of the flux quantum.* The quantization of flux allows us to consider shields which will give an absolutely zero flux state. If we can, by one technique or another, reduce the field to less than one-half a flux quantum through a superconducting shield which is held above its transition temperature, then when the shield is allowed to attain the superconducting state it is energetically favorable for it to exclude all the magnetic flux penetrating it. Small rotational effects change the quantum condition slightly at the 10^{-11} gauss level.

III. SHIELDING AGAINST TIME VARYING FIELDS

The infinite conductivity of the superconductor makes it immediately obvious that it will be a valuable tool to use to shield against time varying fields. We must examine the nature of the shielding currents in order to understand the limitations of the superconductor.

We have already discussed the fact that currents will flow in the superconductor to keep the magnetic field constant. The London equations predict that in the stationary state the magnetic field will decrease exponentially from the surface of the material, the decrease being due to the screening currents. The characteristic length for the decrease is between 370 Å and 500 Å in lead and the currents flow in a surface layer of approximately this thickness. The internal electric field is zero due to the infinite conductivity.

For low frequencies then we expect the screening currents to flow in a surface sheath of approximately 500 Å thickness and to exclude the field from the material inside the sheath. Of course any cavity inside the material will then be perfectly shielded from changing fields outside. Thus a superconductor which is many penetration depths thick should be a much more efficient shield than a normal metal against time varying fields. The frequency at which the shielding by a superconductor becomes no better than that from a normal metal will be the point at which the normal skin depth becomes roughly comparable to the penetration depth. This is somewhat above microwave frequencies.

In this discussion we have implicitly assumed that we are dealing with Type I superconductors and hence will be restricted to
----
* This assumes the path lies in the superconductor where there are no currents. If there are currents along the path the quantity which is quantized is the fluxoid, a quantity which includes the line integral of the current density. (11)

operation in fields which are less than the critical field. We do not need to adhere strictly to this limitation.

If we wish to shield against time varying magnetic fields in the presence of a steady d.c. magnetic field we may use a Type II superconductor as a shield. The Type II shield is kept in the normal state while the d.c. field is applied and then is allowed to become superconducting, trapping the steady field. Since a Type II superconductor still maintains zero resistance while allowing flux to penetrate it, it will still shield against time varying fields that are not strong enough to move the trapped magnetic flux. For this application then we require a superconductor which strongly pins the trapped flux. This dynamic shielding has been used in several experiments which require great field stability in a region of relatively high magnetic field.

IV. SHIELDING AGAINST D.C. FIELDS

We have discussed shielding against time varying fields because the only property of a superconductor which is required is the infinite conductivity. We must examine the Meissner effect and flux quantization to understand how to use a superconductor to shield against a constant field.

We know already that a perfect Meissner effect is never observed. Some magnetic flux remains trapped in the superconductor in individual quanta. We have a picture of magnetic flux trapped in a normal region of the superconductor, that region being held normal by currents flowing around it. The current vortex contains an integral number of flux quanta. Thus the field is everywhere zero in the superconductor except in the region of the normal core. The trapped flux lines cannot easily be moved in the superconductor because of their tendency to become pinned at grain boundaries or dislocations. It is the existence of these pinning sites which prevents a perfect Meissner effect. The pinning sites can be reduced in number by using very pure material and by careful annealing of the shield after fabrication. Our experiments still show a Meissner effect which is far from complete even after these precautions. Means must be devised to decrease the amount of flux which the Meissner effect must remove from the material. The ambient field should be reduced to as low a value as possible before the shield is made superconducting; it may then be raised to any value less than the critical field after the shield is completely superconducting.

Some of the first work on superconducting shielding was done by Vant-Hull and Mercereau (14) and by Deaver and Goree (5). They attempted to decrease the field penetrating a superconducting cylinder by spinning the cylinder rapidly while it was cooled through its superconducting transition. Any ambient magnetic fields which do not have axial symmetry with respect to the cylinder will, when seen from the frame of the rotating cylinder, appear to be a time varying field of frequency ω, where ω is the rotation frequency. Eddy currents will therefore be set up in the rotating cylinder to counteract these time varying fields.

If the cylinder is started rotating while warmed above its superconducting transition temperature these eddy currents will tend to exclude or screen the resulting time varying field but because of the finite conductivity will not screen it completely and will decay if the rotation is stopped. Measurements of the electrical resistivity of lead show that the resistivity of the metal drops abruptly above the transition temperature becoming very small before the metal becomes superconducting. Thus if the cylinder is slowly cooled while rotating the eddy currents will approach those which would be induced if the material had zero resistance, i.e., they will approach the superconducting screening currents.

As the cylinder makes the transition to the superconducting state there will be very little flux penetrating the cylinder since it has already been excluded by the eddy currents. Thus the density of trapped flux lines should be much lower than it originally would have been.

It should be noted that this technique results in a field in the cylinder which has small transverse components. It will not shield against a field which has axial symmetry with respect to the axis of rotation. It thus results in a very uniform field along the axis of the cylinder. If this component is also desired to be small it is necessary to reduce it by using supplementary external coils or high permeability magnetic shielding. This must be done while the shield is being cooled because once it is superconducting the flux is trapped in the cylinder and cannot be removed without driving the cylinder normal.

Deaver and Goree have also measured the attenuation of an external axial field due to a cylindrical shield. They found that the axial magnetic field decreased as:

$$H_z = H_0 \, e^{-\frac{3.4z}{r}}$$

as they went down into the cylinder. This value is in fair agreement with the theoretically predicted exponential constant of 3.8 which is obtained by treating the superconducting shield as a substance which is completely impermeable to the applied magnetic field. The transverse component of an external field should fall off as

$$H_t = H_0 \, e^{-\frac{1.8z}{r}}$$

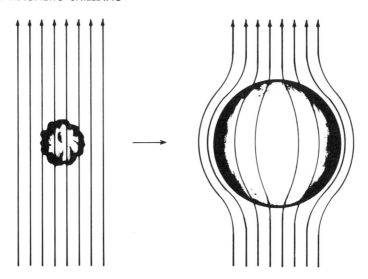

Figure 2. A Folded Superconducting Bladder decreases the field inside when expanded.

and qualitative agreement with this prediction has been obtained by both Deaver and Goree and in our measurements.

We have mentioned before that the Meissner effect has never been seen to be perfect. Bol (1) discovered that the amount of flux excluded from metal cylinders which were being spun depended on such things as the strain in the metal. He found that this strain dependence appeared even when the annealed cylinders were not spun at all until reaching helium temperatures and hence were probably due to the strain induced by the centrifugal forces of spinning.

Because of these measurements we see that it is probably not sufficient to depend on the Meissner effect itself to obtain an extremely low field. However, there are other schemes which utilize the tendency of a superconductor to exclude flux and the quantization of magnetic flux in a superconductor to reduce the field, in theory being capable of reducing the field to zero.

One such method is illustrated in Figure 2. If a folded flexible superconducting bladder is cooled in a given magnetic field it will exclude some of the flux incident on the bladder and trap the remainder. If the bladder is then unfolded the density of flux lines inside the bladder can be much lower, hence the field is lower. If another bladder is then inserted inside the first and cooled slowly enough it should trap less flux than the first because it was cooled in a lower magnetic field. The second bladder can then be

expanded, thus decreasing the field still further and a third
bladder inserted inside the second, and so on. Thus in principle,
it should be possible to decrease the field to an arbitrarily small
value.

There will come the point where the field on an unopened,
nonsuperconducting bladder will be low enough that the flux linked
by the bladder is less than one-half of a flux unit. At this point,
when the bladder becomes superconducting it is energetically
favorable for it to exclude all the flux and thus yield a zero
magnetic field inside.

The first published work on the use of superconducting bladders
was that of Brown.(2) He used cylindrical bladders made of lead
foil and folded so that upon opening they would decrease the field
along the axis of the cylinder. The field transverse to the cylinder
axis was decreased by a geometrical factor depending on the folding
pattern. Brown obtained field reductions by a factor of 500 in
each bladder expansion but did not check the field below 10^{-6} gauss
because his flux gate magnetometer would not respond to lower fields.

V. THE JOSEPHSON EFFECT MAGNETOMETER

As we see from Brown's work our ability to shield against d.c.
fields can approach the limit of our ability to measure the fields
with conventional magnetometers. The Josephson effect magnetometers
however are a new generation of sensitive devices which have been
shown to measure fields down to 10^{-10} gauss. (9, 12, 13)

The Josephson effect is the tunneling of electron pairs from
one superconductor to another through a nonsuperconducting barrier
between them. The phase of the electrons' wave function is affected
by the magnetic field through which the electrons move. In a junction
there can be little shielding of the field by surface currents so
that all of the electrons that participate in the tunneling are
affected by the field. The result of this is that the maximum
tunneling current which can be passed through a single junction is
a periodic function of the applied field to the junction. As the
magnetic field is increased this maximum current, the critical
current, rises through successive maxima and minima. The field
difference between two successive minima corresponds to the addition
of one more flux quantum to the junction. Because an individual
junction is quite small this periodicity corresponds to field differences in the milli-gauss range.

A magnetometer can be made by putting two junctions in a
superconducting loop. The addition of the requirement that the flux
also be quantized around the loop results in the critical current
being periodic in the flux contained in the loop. Since the area

of the loop can be orders of magnitude larger than the area of a single junction the corresponding field sensitivity will be much greater. (6, 16)

Magnetometers are constructed from loops containing double point contacts by arranging a current source to pass through the loop. The current source can be either a.c. or d.c. The voltage across the loop is monitored with a second pair of leads. When the critical current is reached, a voltage will appear. If now the field on the loop is modulated with a small a.c. component and the bias is adjusted to the critical current, an a.c. voltage will appear across the voltage leads, the magnitude of which will depend on how near to the critical value the current is set. Clearly it will also depend on the field through the loop since the critical current is a function of the field. The a.c. voltage can be read with a synchronous detector (lock-in detector) and the output of the lock-in fed back to keep the field through the loop a constant. In this configuration the feedback current is a direct measure of the amount that the flux changed through the loop.(8)

Our magnetometer is what is conventionally known as the double point contact magnetometer. Two Nb foils are pressed around a cylindrical rod so that when they are placed together the depressed parts form a cylinder. A small piece of mylar or kapton is placed between the foils so that they are not in electrical contact. The current and voltage leads are spot welded to the foils. Two sharpened Nb-Zr wires are driven through the foils, one on each side of the cylinder. These form the contact between the foils and the Josephson tunneling occurs through them. These points can be adjusted from outside the dewar when it is at liquid helium temperatures. A modulation coil provides the modulation field to the foils. A superconducting coil is placed inside the cylinder so that any current in the coil will appear as flux in the cylinder and hence will be observed as a change in the reading of the lock-in detector.

This coil is connected to the secondary of a superconducting transformer. The primary of the transformer is connected to a flip coil of approximately 1" diameter. Thus any change in field through the flip coil is measured by the Josephson device. A calibration coil is also wound with the flip coil so that a known current can be passed through the calibration coil to check field sensitivity. This arrangement allows the magnetometer to be treated as a black box which is calibrated by the calibration coil. The flip coil can be rotated through more than 360°.

The entire magnetometer is constructed of non-magnetic materials. All metals are aluminum alloys except for the superconducting foils and wires and a few copper wires. Nylon and Delrin are used for electrical isolation. The entire magnetometer fits into an aluminum can which can be evacuated so that it is not in contact with the liquid helium.

Mercury batteries are used for current and voltage sources. A PAR HR-8 phase sensitive detector is used for the detector and supplies the modulation current. All leads which can couple to the low temperature region are connected through a π-type low pass filter to avoid pick up by the leads. These filters are in an aluminum box at the room temperature end of the apparatus. The sensitivity of the apparatus is 10^{-9} gauss with a 1 second response time.

VI. EXPERIMENTS IN SUPERCONDUCTING SHIELDING

We have performed a series of experiments which were designed to obtain as low a magnetic field as possible. We believe we have obtained a field as low as 6×10^{-8} gauss, over a large region, though as will be seen below, this is subject to various interpretations. We do believe that our measurements demonstrate that superconductors offer the possibility of obtaining a truly zero magnetic field and offer shielding factors of at least 10^8 against variations in electric and magnetic fields outside the shield.

A partially folded balloon is shown below. It is made of lead foil .0025" thick. The foil is soldered without flux to make the

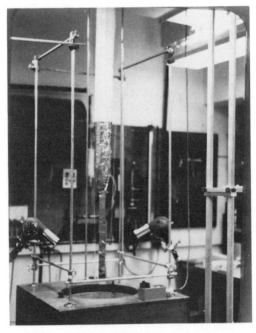

Figure 3. A partially opened lead bladder hangs over the dewar opening. The flexible leader at the top is cotton cloth. The bladder is manipulated by nylon fish line.

balloon. Due to the folding of the balloon we expect the horizontal component of the field to be greater in the direction through the unfolded face. The field plot in this direction is shown in Figure 4 while the perpendicular direction is shown in Figure 5. This balloon was cooled in a field of approximately 10^{-4} gauss which was the residual field inside a double μ metal shield which surrounded the dewar. The field reduction is apparent.

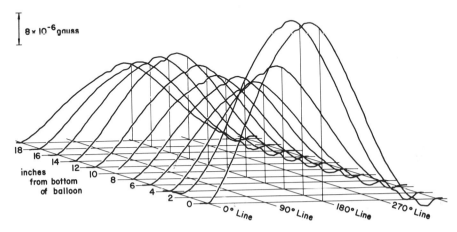

Figure 4. The output of the flip coil vs. rotation at various positions in an unfolded balloon. The balloon was not heat flushed. The field is horizontal and approximately 8×10^{-6} gauss at the point of lowest field. The coil was started when its plane was parallel to the unfolded face of the bladder.

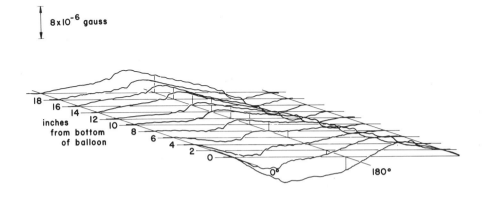

Figure 5. Same as Figure 4 but with the coil started with its plane parallel to the previously folded sides.

We measured the shielding factor by passing a complex of superconducting coils along the outside of the balloon and examining the response of the magnetometer. A.C. and d.c. fields of up to 50 gauss could be induced at the outer surface of the lead. We found that no signal could be detected inside the shield except in the case where there was a pin hole in the lead. At that point careful examination showed that the field reaching the inside of the shield was greatly attenuated because the nature of the shield is such that any flux line entering the shield through a hole must also go out through the same hole. For a shield without holes we were unable to detect, at the 10^{-7} gauss level, any effect of a 50 gauss field outside. A 50 gauss field did not measurably change the field configuration in the bladder.

The field inside a superconducting shield can be further reduced by "heat flushing," a technique which is designed to reduce the amount of flux which becomes trapped in a superconducting balloon. If, as we cool a folded superconducting bladder through the transition, we keep the area of the balloon which is in the vicinity of the transition temperature small then it will take a large field to create a flux unit in this intermediate state region. The small

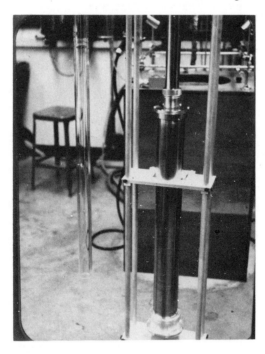

Figure 6. Glass tube which replaced the magnetometer when heat flushing. Beside the glass tube is the magnetometer top assembly. The magnetometer slides through the O-ring tube seal.

transition region is then moved over the entire balloon, flushing out the flux. Some degree of heat flushing can even be obtained by transferring the first helium very slowly into the dewar containing the balloon.

We have used heat flushing to obtain our lowest fields. We constructed a length of double-walled pipe where the space between the walls could be evacuated. This pipe may be observed in Figure 6. A folded superconducting bladder was placed inside the pipe which was then inserted into a previously expanded balloon. The pressure of the helium gas in the pipe could be controlled to allow the level of the helium to rise in the pipe, thus cooling the unfolded balloon from the bottom. We have measured the temperature profile and find that a very sharp gradient, on the order of several degrees per cm, can be maintained across the unfolded balloon. The level difference was observed with a differential oil manometer. The level is allowed to rise over the period of several hours. The rate of rise must be made slow to avoid the generation of trapped flux due to thermoelectric currents induced by the changing temperature difference in the lead.

The field plot inside a superconducting balloon which had been heat flushed is shown in Figure 7. At first sight there appears to be no improvement. However we note that the field in the balloon is considerably higher near the bottom of the balloon than higher up, leading us to the conclusion that we either had an excessive cooling rate when the balloon first started cooling, or that the strains caused by the Delrin block which attached the balloon at the bottom pinned the flux strongly. A second balloon was heat flushed inside

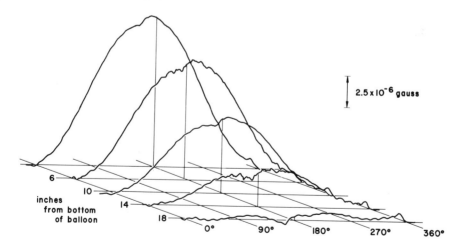

Figure 7. Field after heat flushing. The flux was strongly pinned at the bottom.

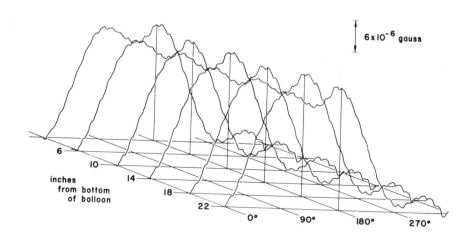

Figure 8. Field at various heights in the balloon after cleaning the magnetometer.

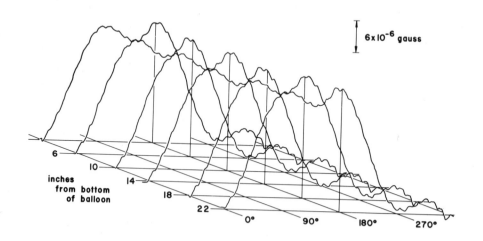

Figure 9. Field in the balloon after rotating the magnetometer 180° in the balloon. If the field rotates with the magnetometer it should be identical to Figure 8. Subtraction of Figure 8 and Figure 9 gives the field in the balloon < 6×10^{-8} gauss.

the first and its field appeared higher than the one it was cooled in. Because of its homogeneity we decided the abnormally high field was probably due to ferromagnetic contamination of the magnetometer and when it was disassembled a small piece of dirt was found. The dirt was about the size of a printed period. When the field was measured again we obtained the plot of Figure 8. Note that, though high, the field is remarkably uniform in the balloon. We concluded that the magnetometer was still contaminated with some dirt which we could not find. Since the field is horizontal we can tell whether it is due to trapped flux in the bladder or to contamination in the magnetometer by rotating the magnetometer and seeing if the field changes. That field caused by the magnetometer will rotate with it while the rest will not. The field profile obtained on rotation of 180° is shown in Figure 9. The field due to the balloon can be obtained by subtracting the two plots. We estimate that the largest field in the balloon is 6×10^{-8} gauss ($\underline{+6 \times 10^{-8}}$). Larger deviations at points in the balloon might be due to trapped flux but are also of a size to be consistent with the field which would be observed inside a non-cylindrical shield if there were a magnetic impurity on the magnetometer.

These measurements indicate the need for extreme cleanliness in the performance of experiments to obtain very low fields. The fine structure which can be noted in all of the field plots is apparently due to a microscopic ferromagnetic inclusion, possibly in the worm gear which drives the flip coil. We note that there are 28 oscillations in the fine structure and there were 28 turns of the worm required for a 360° rotation of the flip coil. This is only compatible with an impurity in the magnetometer closely coupled to the shaft which turned the worm gear.

We are presently constructing a new apparatus in which every effort has been made to exclude contamination. The magnetometer parts have been chemically cleaned and the magnetometer constructed in a clean bench. The balloons have also been made in a clean room. This apparatus will be used in experiments to determine the limits of the techniques of superconducting shielding.

These experiments indicate that it is possible to use superconducting shielding to obtain very low electric and magnetic fields. The low electric fields come about as a result of the perfect conductivity of the superconductor and the very low magnetic fields because of the quantization of magnetic flux. It should be emphasized however that even without striving for low magnetic fields the superconducting shields make sensitive electrical measurements at low temperatures extremely easy. Pick-up and interference are eliminated if the sensor is surrounded by a superconducting shield. The noise level of our magnetometer over a 6-hour period is shown in Figure 10.

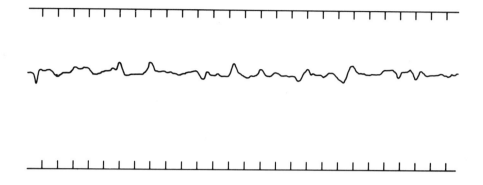

Figure 10. Noise level of our d.c. point contact magnetometer. Full scale is approximately 10^{-7} gauss and the time shown is approximately 6 hours. The magnetic noise level outside the superconducting shield is approximately a factor of 10^6 higher.

VII. EXPERIMENTS UTILIZING SUPERCONDUCTING SHIELDING

A. The He^3 Electric Dipole Moment

One experiment which depends on the zero magnetic field which can be obtained by superconducting shielding is the experiment to observe a static electric dipole moment in the He^3 nucleus and thus check time reversal invariances. If a sample of He^3 liquid should have its nuclear moments aligned and then be placed inside a superconducting shield from which the last quantum of flux has been excluded, the nuclear moments should not precess, there being no external fields to exert a torque on the nuclear angular momentum. If the nucleus has an electric dipole moment then an electric field applied to the sample will cause the nuclear spins to precess. The spin precession can be monitored by measuring the magnetization of the sample since the magnetic moments will precess with the spins.

This experiment, which has been pursued in Fairbank's group for several years, is extremely difficult and would be almost impossible without superconducting shielding. The He^3 nucleus precesses 2×10^4 radians/sec/gauss in a magnetic field and hence any magnetic field would mask the very small precession which would be expected to be due to an electric dipole moment in any electric field. If a shield from which the last quantum of flux has been excluded can be placed around the sample the apparatus under construction may ultimately be capable of detecting an electric dipole moment of 10^{-24}e cm which is of the order which would be expected if time reversal invariance were violated in the weak interactions.

The most recent results from this experiment are that a sample of He^3, aligned by optical pumping, has been condensed at low temperatures inside a superconducting shield which had an inhomogeneous field of approximately 10^{-6} gauss inside a superconducting shield. The magnetization was monitored with a superconducting magnetometer of the type discussed by Mercereau and Nisenoff (12, 13). The He^3 magnetization was not observable for a long period of time because of the dephasing of the spins due to the field inhomogeneities. The work on this experiment by I. Bass, M. Taber, and T. Jach is continuing.

B. The Measurement of Magnetic Fields Due to Biological Activity

It has already been demonstrated that there is a magnetic field created when a muscle contracts. The magnetic fields associated with cardiac activity have been observed (3) and, by using an averaging computer triggered by the EEG recording, a magnetic field associated with the α rhythm of the brain has been seen. The magnetic fields associated with these biological activities are on the order of 10^{-7} gauss for cardiac activity and 10^{-9} gauss for activity in the brain. The frequencies of this activity range from approximately 1 hz up to 100 hz.

The difficulties of measuring such small fields at these low frequencies are obvious and the difficulties associated with using conventional shielding are well known. For instance, the skin depth in aluminum is on the order of 1 cm, so high permeability materials offer the only possibility of shielding by conventional means. Yet these materials are extremely difficult to use below 10^{-5} gauss. On the other hand a single superconducting shield can offer perfect shielding against time varying fields, both electric and magnetic.

Work is presently underway to construct a region which is large enough to hold a man. This work, under Fairbank at Stanford, is being pursued by J. Opfer and J. Madey. They do not require zero field but only as near perfect shielding against time varying fields as possible. It is also desirable to avoid field gradients as much as possible so that motion of the subject or the magnetometer does not result in a false signal at the output.

C. Experiments in Gravitation

The recently announced discovery of gravitational radiation by

Weber (15) has opened vast questions in astrophysics. The measurements are very difficult, involving the detection of strains in aluminum bars 5 feet in length when the motion causing the strain corresponds to a motion of the bar's end of 10^{-14} cm. The shielding problems associated with the detection of such small strains are extreme. For instance, not only could electromagnetic energy at the resonant frequency of the detector cause a spurious signal but, by the magneto-acoustic effect, so could a steadily changing magnetic field. The signals which Weber detects are of the same magnitude as the mean thermal strains in the bar; the signal-to-noise ratio of his best signals is no better than 5:1 and is limited by the 300°K thermal vibrations of his detector.

We are now constructing a gravitational wave detector which will be cooled to below liquid helium temperatures. The purpose of the cooling is to decrease the thermal noise in the bar so as to improve the signal-to-noise ratio and increase the sensitivity. The low temperature environment also allows us to support the detector on a superconducting magnet, a technique which should not damp any of the longitudinal modes of the detector and which, therefore, should further decrease the system noise. With the low temperature environment we also have the opportunity to shield the experiment with superconducting foil to eliminate any chance of electromagnetic pick-up or excitation.

The motion of the massive detector will be monitored with a superconducting accelerometer. This accelerometer, by using superconductors as the active elements, shows promise of being very stable and noise free. Moreover by monitoring the acceleration of the end of the bar one is able to look for all of the normal modes of the bar and hence to do a crude spectral analysis of the driving gravitational wave.

The accelerometer consists of a superconducting mass supported on a persistent current magnetic spring of long time constant. Motion of the case which contains the magnet will cause the position of the mass to change with respect to the magnet. This change of position will be reflected as a change in the magnet inductance. Clearly the accelerometer depends on the constant magnetic field support and hence on adequate superconducting shielding to insure its constancy.

This work is being pursued jointly at LSU and Stanford. Our work at LSU is being done by P.B. Pipes with assistance from D. Baker and C. Schueler, while the Stanford members of the project are W.M. Fairbank, M.S. McAshan, S. Boughn and R. Taber. The construction of the gravitational radiation experiment is funded by the National Science Foundation.

VIII. CONCLUSIONS

Superconducting shielding offers the possibility of extremely high attenuation (shielding factors of 10^7) of time varying external fields from very low frequencies up to frequencies in the microwave region. We have demonstrated that it is also possible to obtain exclusion of steady external fields to the extent that a region 4" in diameter and several feet long had a field of less than 6×10^{-8} gauss. This type of shielding is not difficult to use in most low temperature apparatus.

IX. ACKNOWLEDGMENTS

We wish to express thanks to the great many people who have contributed their time and talent to the work reported here. Much of the work was a group project. In particular we wish to thank W. M. Fairbank for his support and encouragement. A great deal of the ground work for the magnetometer design was done by A. Hebard. D. K. Rose gave valuable help at critical points.

REFERENCES

(1) M. Bol, Ph.D Thesis, Stanford University (1965)

(2) R. E. Brown, Rev. Sci. Instr. 39, 547 (1968)

(3) D. Cohen, E. A. Edelsack, and J. E. Zimmerman, Appl. Phys. Letters 16, 278 (1970)

(4) B. S. Deaver and W. M. Fairbank, Phys. Rev. Letters 7, 43 (1961)

(5) B. S. Deaver and W. S. Goree, Rev. Sci. Instr. 38, 311 (1967)

(6) R. deBruyn Ouboter and A. Th. A. M. deWaele, Revue de Physique Applique 5, 41 (1970)

(7) R. Doll and M. Naubauer, Phys. Rev. Letters 7, 51 (1961)

(8) R. C. Forgacs and A. Warnick, Trans. IEEE IM-15, 113 (1966)

(9) W. S. Goree, Revue de Physique Applique 5, 3 (1970)

(10) W. O. Hamilton, Revue de Physique Applique 5, 41 (1970)

(11) F. London, Superfluids Vol. I, Dover Publications, New York (1961)

(12) J. E. Mercereau, Revue de Physique Applique 5, 13 (1970)

(13) M. Nisenoff, Revue de Physique Applique 5, 21 (1970)

(14) C. C. Vant-Hull and J. E. Mercereau, Rev. Sci. Instr. 34, 1238 (1963)

(15) J. Weber, Phys. Rev. Letters 22, 1320 (1969); 24, 276 (1970)

(16) J. E. Zimmerman and A. H. Silver, Phys. Rev. 141, 367 (1966)

SUPERCONDUCTIVE COMPUTER DEVICES

J. Matisoo

IBM Thomas J. Watson Research Center

Yorktown Heights, N.Y. 10598

I. Introduction

Not many devices have found use in digital computers either for reasons of technology or economics. Very early machines utilized mechanical relays from which both logic and memory circuits were built. These were quickly replaced by vacuum tubes and then transistors. At the same time, special devices which could store information but could not in general be used for logic operations evolved. Among these are Williams tubes, magnetic drums and finally magnetic cores. As the need for storing more and more information arose, magnetic tapes and discs were developed.(1)

The transistor and the magnetic core, the dominant computer devices of the present day, both date from the early fifties. Superconductive devices date from the same period. In 1956, Buck proposed a device which he called a cryotron.(2) It consists of a niobium wire wrapped around a tantalum wire. When no current flows in the niobium loop, the tantalum wire is superconducting and, therefore, has zero resistance. When current through the niobium wire generates a field sufficiently large to drive the tantalum wire normal, it has its normal state resistance. The device is thus a switch with an infinitely large resistance ratio. When used in circuits, it has properties similar to those of a mechanical relay.

In its wire-wound form the cryotron was never a practical device, because its speed of operation in any computer circuit was slower (ms) than that of even crude room temperature devices, certainly much slower than that of early transistors. A period of device development followed. For transistors, the development was intensive, and progress in transistor performance and miniaturization has been remarkable, having already progressed through the stages of integrated circuits to large scale integration.[3] For superconductive devices, the development effort was not as strong and no device has yet come into commercial use. To be fair, however, it should be pointed out that 4°K operation automatically places requirements on any device or computer system. First, if the performance of a superconducting device is comparable with that of a room temperature device, the latter will be preferred for convenience. Secondly, if device costs are comparable, the system cost for superconducting devices must include the cost of refrigeration to operate at 4°K. To be competitive, therefore, the developers of superconductive devices were forced to consider systems with a very large number ($\sim 10^7 - 10^8$) of devices in which the refrigerator cost would be negligible. To make devices in such large numbers, they adopted large scale integration techniques. But they were before their time by perhaps as much as ten years, and most of these efforts failed. The semiconductor technologists are facing similar problems today.

After considering some necessary and desirable properties of computer devices in general, we shall discuss the basic ideas underlying all superconductive computer devices. We shall briefly discuss thin film cryotrons, simple logic and memory circuits, and proposed or existing systems. The Josephson tunnel junction devices, which are more recent and have considerable potential, will receive somewhat more detailed discussion.

II. <u>Basic Ideas</u> (See Reference 1)

All computer devices in use are devices which possess two distinguishably different states and can be switched controllably between these two states. There is nothing magic about two states, or binary operation, but the relative ease with which one can distinguish without error the difference between two states rather than say, ten, has made binary universal. These devices are combined into circuits which store information (memory) or which perform arithmetic operations on this information (logic). All arithmetic functions can be carried out with "AND", "OR" and "NOT" logical circuits. Frequently, logic devices are distinguished as a class from memory devices even though all logic devices can be used to build circuits which store information. This differentiation is made because the requirements on these generic types of devices are different. Usually a computer requires many more memory devices than logic devices. This places a premium on cost. Logic devices, however, must be fast in operation, so that the computer can perform many operations in a second. Further, these devices should be active; i.e., possess gain so that one device can

SUPERCONDUCTIVE COMPUTER DEVICES

control another without intermediate amplification. The devices should be small and dissipate little power so that they can be packed close together. This not only makes computers small, but because the electrical paths between devices are short, it also makes them fast. Thus, requirements for a logic device are: two states; fast switching between these; active; small; cost, although important, is not as significant as for memory devices. Memory devices should posses two states, which if possible should be stable with respect to removal of electrical power; switch rapidly between states; very inexpensive; small, dissipate very little power. They may be passive.

For reference, we briefly discuss these characteristics of the current standard devices, the transistor and the magnetic core.

In the transistor circuit, a "state" corresponds to a voltage level at some point. For example, one state could be defined as "0" if the collector voltage is +1V and as "1" if it is +1.5V. (Any number of configurations are possible). Switching between these two states might be initiated by altering the base to emitter voltage. In high performance circuits, this might be substantially complete in 3 or 4 ns. The transistor is, of course, an active device; i.e., in appropriate configurations, it exhibits power gain. It can drive a number of other transistors. They can be made quite small, having an active area of perhaps a few mil^2. The operating power levels tend to be rather large ($\sim$50mW.)

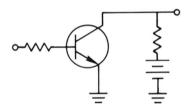

Figure 1 - A transistor switch.

The magnetic core has two stable states in the two possible directions of the remanent magnetization. Wires strung through the core generate magnetic fields which "write" the core magnetization in one direction or the other. Switching speed depends on the size of the core and how hard it is driven. Small cores ($\sim$10 mil diameter) can be switched in a few hundred nanoseconds. The core is passive and retains information in the absence of electrical power.

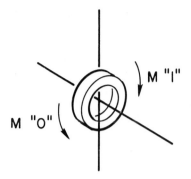

Figure 2 - A magnetic core.

The performance and characteristics of any superconductive device must be compared with these standards.

All superconducting computer devices are based on two ideas - persistent currents, and current steering. Persistent currents are used to store information and logical operations are performed by diverting currents from one superconducting path to another. These ideas can be simply illustrated with reference to Fig. 3. A current source supplies current I_o to a superconducting circuit consisting of two equal parallel legs both of inductance L. Thus current $I_o/2$ flows in each branch. If now by some means we introduce resistance into leg 1, the current will divert to leg r, so that after transients have died out, we have $I_1 = 0$ and $I_r = I_o$. This is obvious so long as $R_1 \neq 0$. It also remains true if we now let $R_1 \to 0$;

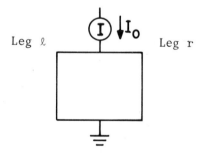

Figure 3 - The basic superconductive logic and memory circuit.

i.e., we let the left leg revert to the superconducting state. This is so because the flux in the loop cannot change. We can, if we like, steer I_o to the left leg by introducing $R_r \neq 0$ so that $I_1 = I_o$ and $I_r = 0$. We can establish a circulating persistent current in the loop by altering the sequence of events slightly. Suppose we make $R_1 \neq 0$ so that $I_1 = 0$ and $I_r = I_o$. If after we let $R_1 \to 0$ we turn off the current I_o, we find a clockwise circulating persistent current, $I_o/2$ (The simplest way to see this is to use superposition). Had $I_r = I_o$ and $I_1 = 0$ the resulting persistent current would have had a counterclockwise circulation. There are simple ways in which the presence or absence of currents and the polarity of persistent currents can be sensed. Because currents in a superconducting loop do not decay the stored information is permanent. Thus, one has the basic elements for information storage and processing. The remainder is detail. This detail, however, can have a very profound effect on the utility of a given device as we shall see in the following discussion.

III. <u>Thin Film Cryotrons</u>

The simplest device is the thin film cryotron. Figure 4 shows the geometry of the cross film cryotron, the earliest of these devices.[4] It consists of a Sn "gate" and a Pb "control" which are electrically insulated from one another by silicon oxide, for example. These lines overlay, but are insulated from, a superconducting sheet which in turn has been deposited on a substrate. Two kinds of fabrication techniques have been employed. Either all metal films are vacuum deposited through metal masks or photolithographic techniques common in semiconductor fabrication may be used to delineate the vacuum deposited metal lines. The width of the lines may be as small as a few mils or as large as twenty. Film thicknesses are usually a few thousand Ångstroms.

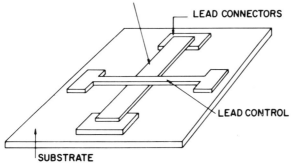

Figure 4 - Thin film, cross-control cryotron
(from J. Matisoo, Proc. IEEE <u>55</u>, 172 (1967))

This device is operated at T = 3.5°K, just below the transition temperature of tin, (T_c = 3.7°K) the idea being to drive the tin "gate" film normal by a magnetic field resulting from current in the Pb "control". The Pb (T_c = 7.2°K) always remains superconducting. In this device, therefore, the resistance which is inserted into the leg to steer current into a parallel superconducting path is the normal state resistance of the tin film under the control. The gate can be driven normal by combinations of gate current I_g and control current I_c. A plot of the locus of points in the I_g-I_c plane separates the superconducting and normal phases of the gate and is referred to as the "gain curve". A typical curve is shown in Fig. 5. Its significance lies in that the gate current for one cryotron in circuits frequently becomes the control current of another. Thus, for any given device, the control current must be smaller than the gate current, otherwise the gate will switch with no control current applied. Gain is defined as the ratio of the critical gate current intercept to the critical control current intercept. This just turns out to be $G = W_g/W_c$ (essentially) where W_g is the width of the gate and W_c is the width of the control. Why this is so is physically clear. At a given temperature, the Sn film has a critical field, H_c, and a critical current density j_c. The critical current is then proportional to W_g. The control field however, is proportional to $1/W_c$. Thus the critical control current is α W_c and the ratio of gate to control critical currents is α W_g/W_c. Thus by choosing this ratio to be greater than one gives a device which can control another without intermediate current amplification.

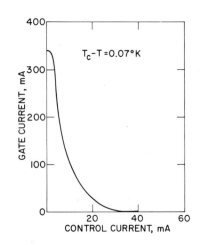

Figure 5 - "gain" curve for cross-control cryotron.
(after V. L. Newhouse, Applied Superconductivity, p. 198, John Wiley & Sons, New York, 1964.)

Two characteristic times govern the dynamic operation of the cryotron. One is the length of time required to introduce resistance into the gate; i.e., the time required for a superconducting to normal phase transition. The other is the circuit time constant associated with current steering; i.e., L/R. The L depends upon the loop dimensions, the R upon gate dimensions and the normal state resistivity of the gate metal, Sn. One can define a minimum inductance (associated with the device) regardless of the circuit into which it is placed; i.e., the inductance of the control crossing the gate.

Doing the R and L calculations (R = $\rho \frac{W_c}{W_g d}$; d - thickness of the gate film and L = $\mu_o \frac{W_g t}{W_c}$, where t is the separation of the control from the ground plane) we find that

$$\tau_{min} = (L/R)_{min} = \frac{\mu_o}{\rho} td \left(\frac{W_g}{W_c}\right)^2$$

Typical numbers might give ∼10ns for $(L/R)_{min}$. In a circuit, this is likely to be ∼100 ns or more. We note that τ is proportional to (W_g/W_c) and that the device gain, G, is also proportional to W_g/W_c. Thus the desire for large G and small τ conflict.

The dynamics of the phase transition turned out to be more complicated than initially supposed. It was assumed that the normal phase nucleated uniformly on the surface of the gate film and propagated through the thickness (∼5000Å), delayed only by eddy current damping in the normal region. On this basis, the gate should have its full resistance in ∼1ns. However, it was experimentally found that regardless of how large a field was applied to drive the gate normal, full resistance took at least 10 to 40 ns to develop.[8] Two effects were found to limit times. In fields with small overdrives (control currents ∼10% larger than required), the latent heat of the transition lowers the operating temperature and increases the critical field to a value which is very close to the applied field; (i.e, the overdrive disappears). The transition time is then governed by thermal time constants which are ∼100ns. Secondly, it was found that the normal phase is not nucleated uniformly over the gate surface, but rather in islands. These propagate laterally and typically take 10 to 40 ns. Thus, the minimum switching times associated with cryotrons are ∼25 ns. In actual circuits the times are much longer, ∼250ns.

Typical cryotron circuits are illustrated in Fig. 6 and 7.[6] Figure 6a shows a flip-flop. Here current can be steered from one leg to the other by control currents and the circuit remembers which control circuit was "on" last. A logical "OR" is performed by circuit 6b. Here current is steered to the right branch whenever one or the other cryotron is "on". A logical "AND" is shown in Fig. 6c. One and the other cryotron must be "on" before current is transferred to the other path. These circuits can be combined to perform more complex functions, by making the right leg a control of another cryotron. A principle followed with all these devices in circuits is that there is always at least one superconductive path between dc terminals and only one path containing controls should be left open to prevent divided or circulating current conditions which can interfere with proper circuit operation.

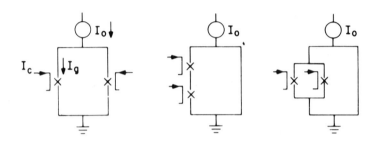

Figure 6 - Typical cryotron logic circuits:
(a) Flip-flop. (b) "OR" circuit. (c) "AND" circuit.

Figure 7 shows a typical memory cell. Writing is accomplished by pulsing both "write" and "write enable" lines, but turning off "write enable" before the "write". This leaves a circulating current in the loop. Reading is done by pulsing "read enable". If current is present in the loop, the "read" current will be diverted elsewhere, if not, it will not be diverted. Many kinds of memory circuit arrangements are possible.

A device variant is the so-called in-line cryotron, in which the control is parallel to and over the gate.[5] This makes $W_g = W_c$ and minimizes the time constant τ, giving a faster device. It has, no static gain. The device gain curve, however, is not symmetric because I_g and I_o fields add and subtract when I_g and I_c are parallel and anti-parallel. Biasing schemes to give gain are thus possible.

Memory cell variants called the "bridge" and "loop" cells are illustrated in Figs. 8a and b.[7] These cells have good tolerances to variations of device parameters in fabrication and to variations in operating conditions.

Other memory schemes deserve mention. One of these is illustrated in Fig. 9.[8] A hole in a superconducting sheet is bridged by a narrow strip superconductor. Information is stored by means

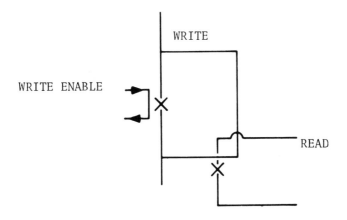

Figure 7 - A possible memory cell configuration.

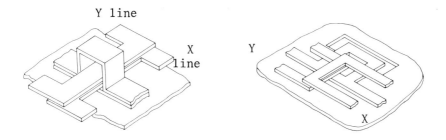

Figure 8 - (a) "Bridge" cell. (b) "Loop" cell.
(after A. R. Sass et al., IEEE Spectrum, 91, July 1967.)

of circulating persistent currents either in the right or left half of the hole. Writing is accomplished by driving the bridge normal with a combination of magnetic fields and current through the bridge itself.

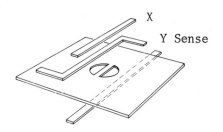

Figure 9 - "Crowe" cell memory.

A variation of this, called the Continuous Film Memory omits the hole.[9] The scheme is illustrated in Fig. 10. It shows perpendicular lead lines over a superconductive indium sheet. The drive currents through the lead lines are chosen so that either separately produces a field below the critical field of indium and magnetic flux will not penetrate. At the intersection, however, field components add and drive the part of the film under the intersection normal, trapping flux which constitutes storage. Reading is done by driving the spot normal with currents of opposite polarity. The change (or no change) in flux produces a voltage (no voltage) which is sensed by the underlying sense line.

No detailed performance characteristics for computers constructed entirely of cryotrons have been given. However, designs for 10^7 bit random access memories with $\sim 1\mu s$ cycle time employing "loop" cells have been described by RCA.[10] Experimental subsystems have been operated.

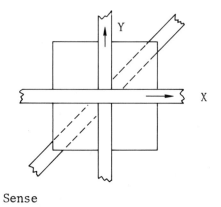

Figure 10 - Continuous Film Memory Cell

SUPERCONDUCTIVE COMPUTER DEVICES

One can ask why no computer systems using these devices have been built even though they possess attractive features, such as simplicity. The likely answer is simple (and has already been given): the device performance is not competitive with that of room temperature devices.

The device which we describe in what follows is sufficiently competitive to warrant another look at superconductive computer possibilities.

IV. Josephson Tunnel Junction Technology[11]

The use of a Josephson tunnel junction as the gate gives a device of superior performance. We review briefly, with the aid of Figs. 11 and 12 the pertinent aspects of the Josephson Effect and illustrate the conditions under which they are observable.[12] Figure 11 shows a "gedanken" experiment. Two pieces of Pb separated by a distance D are connected in a circuit which measures the I-V characteristics of the system as a function of temperature T and spacing D. If T is 300°K and D is a macroscopic distance, 1 cm, say, no current flows through the circuit. If we make D = 50Å

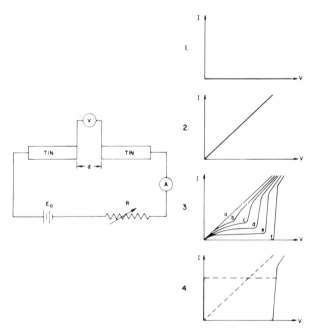

Figure 11 - An experiment illustrating a set of conditions under which the Josephson Effects become observable. 1. T = 300°K, d = 1 cm; 2. T = 300°K, d = 50Å; 3. Temperature decreasing a → f from 7.2°K to 1°K, d = 50°; 4. T = 1°K, d = 25Å.

current flows through the space by electron tunneling. The I-V characteristic is linear. By cooling the metals to 4°K, where Pb is superconducting, the I-V characteristic becomes nonlinear, with little current flowing until the gap voltage, $V_g = 2\Delta/e$ is reached. What has happened is that the current flow below V_g is limited by available single electrons, which are condensing into Cooper pairs and D is such that no superconducting correlations extend across D. (At T = 0, all electrons near the Fermi level have formed pairs and current below V_g is zero.) At V_g, sufficient energy to break pairs is available from the electric field so the current increases rapidily. Finally, when the spacing D is reduced to $\sim 25\text{Å}$, zero-voltage current of limited magnitude appears, in addition to the single electron current. (Of course, the current level here is much greater than at D = 50Å, since tunnel currents decrease exponentially with thickness).

The spacing has now become small enough so that pair correlations extend across the intervening space. The associated energy is appreciably greater than kT, so that the isolated pieces of Pb act very much like a single piece of superconductor. Nevertheless, the superconductivity in region D is weak and electromagnetic potentials can be sustained. In a sense, then, the zero-voltage current is ordinary supercurrent and the maximum value I_J represents the critical current. This is the dc Josephson effect. The value of I_J is a strong function of magnetic fields in the vicinity of D. Figure 12 shows a typical dependence of I_J on fields applied in the plane of an actual junction. The Fraunhofer diffraction pattern arises because the local current depends on the local magnetic field and these can be made to interfere in such a way that the total current I_J varies as shown. Josephson has written down the three basic equations which describe the behavior of these currents extremely well. The field dependence $I_J(H)$ follows directly from these. Depending upon junction geometry and the current density, many other field dependences can be obtained.

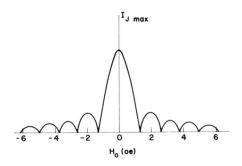

Figure 12 - The typical magnetic field dependence of the dc Josephson current.

From this discussion, it is obvious that a Josephson tunnel junction has two states (V = 0, and V = V_g) and the switching between these two states can be controlled by means of a small magnetic field. The device geometry is thus very much like that of a cryotron in which, however, the gate is not a tin film but a junction. One can ask what benefit is likely? Greatly increased device performance can be expected, simply because in switching the junction, there is no superconducting to normal phase transition. Further, the nonlinearity and hysteresis of the I-V curve should lead to much improved circuit performance. The fields required for switching are oersteds rather than the hundreds of oersteds required to drive a superconducting film normal.

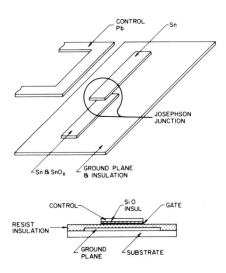

Figure 13 - Josephson Tunneling Cryotron device structure (After Reference 13).

The device structure and fabrication are illustrated in Fig. 13.[13] On an insulating substrate one deposits a lead ground plane, and covers it with silicon oxide insulation. On this, the base electrode of any convenient superconductor such as Pb is evaporated through a mask. The tunneling barrier is formed by oxidizing Pb either thermally or in an oxygen plasma, to an oxide thickness of about 25Å. The junction is completed by evaporating a Pb counterelectrode. To provide magnetic field, a control line of Pb is evaporated above, but insulated from, the junction. An actual device I-V characteristic is shown in Fig. 14. How the zero voltage current I_J depends on H is shown in Fig. 15. The field dependence is rather different from a Fraunhofer pattern because the

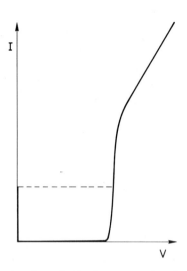

Figure 14 - Device I-V characteristic traced with a current source. The hysteresis is fully developed.

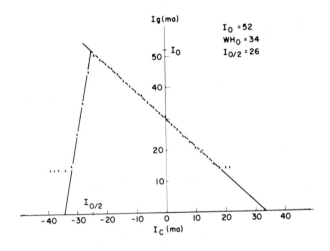

Figure 15 - Device "gain" curve for a particular value of current density and dimensions.

length, in this case, is large compared with the Josephson penetration depth, λ_J. Figure 15 is also the device gain curve. It is clear that if biased to near the maximum I_{max} value by means of an ancillary line, large current gains can be obtained; (i.e., a small ΔI_c leads to a large ΔI_j).

SUPERCONDUCTIVE COMPUTER DEVICES

The junction switching time is determined by the junction capacitance, I_J, and V_g. To a very good approximation, the time required for the junction to undergo the transition from $V = 0$ to $V = V_g$ is just the time required to charge the junction capacitance to V_g with a current source I_J. Since $I = C\, dV/dt$

$$\Delta t \simeq \frac{C V_g}{I_J}$$

For Pb (at 4°K) $V_g = 2.5$ mV. If $I_J = 6.3$ mA and $C \simeq 2 \times 10^{-10}$ f, then $\Delta t \simeq 80$ ps, which is a very short switching time indeed. For an experimental junction having these parameters an 80 ps switching time was expected and found.

To obtain an idea of operating speed in a circuit, consider the current transfer problem in a flip-flop, illustrated in Fig. 3[6]. The current is initially in the left branch and will transfer to the right after the left device switches to the V_g state. The device in the V_g state looks like a battery (because the differential R is very small), so that the time required to transfer current I_J can be calculated from $V = L\, dI/dt$, where V is the gap voltage V_g^J and L is the inductance of the loop. $\Delta t = L I_J/V_g$. If $L = 10^{-10}$ $I_J^g = 10$ mA then $\Delta t = 400$ ps, (the actual situation is more complicated, but it does not greatly alter the argument). These numbers have actually been measured.

The circuitry for memory and for logic is very much the same as has already been described for cryotrons, with vastly improved performance. Again many variations on a theme are possible.

To date, a single memory system proposal based on the tunneling devices has been published. The system envisioned has 30 million bit capacity and a cycle time of 15 ns or 40 ns depending upon the choice of array size.[14]

There are a number of technological problems. The oxide barrier is quite thin, 25Å, or so. Because the tunneling current depends exponentially on the oxide thickness, good control of the thickness is necessary if any reproducibility is to be achieved.

Other technological problems are stability of the structure with respect to thermal cycling and stability with respect to room temperature storage. These could be overcome by proper choice of materials so that expansion coefficients are matched and metallurgical stability assured.

V. Summary

We have described essentially two kinds of superconductive computer devices, the thin film cryotron and the Josephson tunneling cryotron, and have illustrated their use. The thin film cryotron is a relatively slow device. Switching times are about 25 ns at best and circuit time constants are ~250ns. Power dissipation is small, ~100μW in continuous operation. It is an active device. The Josephson tunneling cryotron is a very fast device with ~80ps switching time, and can be operated in circuits in a few hundred picoseconds. Power dissipation is ~10μW in continuous operation. It is an active device. This device has performance which is competitive with room temperature devices and should receive further development.

References

1. See, for example, N. R. Scott, Electronic Computer Technology, McGraw-Hill Book Company, New York, N.Y. 1970.

2. D. A. Buck, Proc. IRE 44, 482 (1956).

3. For a review of Large-Scale Integration, see, for example, F. G. Heath, Scientific American, 222, 22 (1970); for state of the art transistor performance, K. E. Drangeid and R. Sommerhalder, IBM J. Res. Devlop. 14, 82 (1970).

4. V. L. Newhouse and J. W. Bremer, J. Appl. Phys. 30, 1458 (1959).

5. A. E. Brennemaun, J. J. McNichol and D. P. Seraphim, Proc. IEEE 51, 1009 (1963).

6. R. Edwin Johnson in: Handbook of Thin Film Technology, L. I. Maissel and R. Glang, Eds., McGraw-Hill Book Company, New York, New York 1970.

7. A. R. Sass, W. C. Stewart, L. S. Cosentino, IEEE Spectrum 4, 91 (1967).

8. J. W. Crowe, IBM J. Res. Develop. 1, 295 (1957).

9. L. L. Burns, Jr., G. A. Alphonse and G. W. Leck, IRE Trans. EC-10, 438 (1961).

10. R. A. Gange, Electronics 40, 111 (1967).

11. For a review, see, for example, J. Matisoo, IEEE Trans. Magnetics MAG-5, 848 (1969).

12. B. D. Josephson, Advan. Phys. 14, 419 (1965).

13. J. Matisoo, Proc. IEEE $\underline{55}$, 172 (1967).

14. W. Anacker, IEEE Trans. Magnetics $\underline{\text{MAG-5}}$, 968 (1969).

SUPERCONDUCTORS IN THERMOMETRY

J. F. Schooley

National Bureau of Standards

Washington, D. C. 20234

Various properties of superconductors show monotonic temperature dependences, so that in principle they may be used as thermometric parameters. Other properties can be utilized in devices which find convenient application in thermometry. Of the many possible examples, we will examine several which have found some actual use either as thermometers or detectors in thermometry.

HEAT CAPACITY

One of the first uses of superconductivity in thermometry came about in a study of the heat capacity of vanadium by Corak, Goodman, Satterthwaite and Wexler (1). The question at hand was the functional form of the heat capacity below T_c; the authors wished to compare their results with the T^3 relation arising from the two-fluid model of Gorter and Casimir (2), and with an energy gap model proposed by Fröhlich, by Kuper, and by Bardeen (3). The latter involved the exponential relation

$$C_{es}/\gamma T_c = ae^{-bT_c/T}$$

In trying to fit these analytic relations to the vanadium and other heat capacity data, the authors found substantial systematic errors which they ascribed to the temperature scale currently in use, T 48 (4). Consequently, in fitting the heat capacity data they used a scale deliberately chosen to smooth out these systematic errors. As is well known, they found rather better agreement with the exponential "energy gap" relation than with the T^3 two-fluid form, thus helping to point the way to the Bardeen, Cooper,

Schrieffer theory. The experiment demonstrated quite effectively, however, that the heat capacity can serve particularly well to detect lack of smoothness in a temperature scale below T_c. This is due not only to the fact that heat capacity follows an analytic temperature relation, but also to the fact that it involves a differential in temperature. Normal-state heat capacities have been used to examine temperature scales in a similar way, of course, but neither normal-state nor superconductive heat capacity can be used as a primary thermometer.

CRITICAL MAGNETIC FIELD

A second use of superconductivity in thermometry involves the dependence of the critical magnetic field on temperature. In this technique a magnetometer is used to detect the transition of a superconducting sample as a function of applied magnetic field. The value of the critical field so derived is then unique to the temperature at which the measurement was made. If a simple parabolic relation for critical field

$$H_c(T) = H_o (1 - T^2/T_c^2)$$

were obeyed exactly, a one-point temperature calibration would provide immediately a useful thermometer down to about $T_c/4$. In fact, however, the BCS theory predicts a slightly flattened parabola for the critical field curve (5), so that the function

$$D\left(\frac{T}{T_c}\right) = \frac{H_c(T)}{H_c(0)} - \left[1 - \left(\frac{T}{T_c}\right)^2\right]$$

is negative between $T = 0$ and $T = T_c$. Approximate calculations of the critical field curves are possible in certain cases; the data for aluminum nearly fit the isotropic, weak-coupling limit calculation of the BCS model, and the lead data were fitted approximately by a strong-coupling calculation of Swihart, Scalapino, and Wada (6).

Thus the temperature dependence of the critical magnetic field must be measured against a primary thermometer, rather than being derived from theory. Subsequently, the data so evaluated can be used as a secondary (calibrated) thermometer. Harris and Mapother have been able to measure the critical field of Al samples in this way above 0.3 K (7), and recently, Ries and Mapother have suggested that a convenient critical field thermometer based on Al can provide ± 1 mK accuracy between 0.3 K and 1.2 K (8).

SUPERCONDUCTIVE TRANSITION TEMPERATURE

The International Practical Temperature Scale relies upon narrow, reproducible phase changes as defining fixed points (9). For temperatures as low as 27 K, freezing points, boiling points, and triple points provide useful fixed points. However, the hydrogen boiling point at 20 K is difficult to realize experimentally as are the hydrogen triple point at 13 K and the helium boiling point at 4 K. Therefore, there has been considerable interest in evaluating other phenomena as possible cryogenic fixed points.

The superconductive transition offers a convenient possibility, since transition temperatures vary from 20 K to 0.01 K or lower, the transition is easily observed, samples can be made small, high purities are available, and relatively high thermal conductivities enhance the likelihood of temperature equilibrium throughout the system.

The natural width of the superconductive transition in gallium was shown to be small by Gregory (10). Observing the transition in single crystals by means of mutual inductance measurements, he found the width of the gallium transition to be as small as 20 μK, which is sufficiently narrow for use as a thermometric fixed point.

The widths and reproducibilities of Pb, In, Al, Zn, and Cd were examined in a long series of experiments by Soulen and Schooley, in an effort to provide a series of fixed points from 7 K - 0.5 K (11). We found transition widths to be generally less than 1 mK, and the transitions were reproducible to within 1 mK, as measured against a set of germanium resistance thermometers during several thermal cycles. Samples of these superconductors are now being prepared for general use as cryogenic fixed points. In an interesting related experiment, Soulen and Colwell demonstrated that the superconductive transition in indium as measured by mutual inductance coincides with those observed by resistivity and by heat capacity measurements (12).

Thus, it appears that superconductive transitions can serve as thermometric fixed points, and it remains to find suitable transitions at temperatures above 7 K and below 0.5 K.

SUPERCONDUCTIVE JUNCTION TUNNEL CURRENT

Superconductive tunnel junctions form the basis for several thermometric schemes. When a thin oxide layer separates a normal metal from a superconductor, the tunneling current is a strong function of both voltage and temperature. Expressions for the tunneling current were derived by Giaever and Megerle (13); in principle then, one could simply measure the tunneling current as a thermometric parameter.

As given, for example, by Douglass and Falicov (14), the ratio of the tunnel current at V = 0 for one metal superconducting to that for both metals normal has the exponential temperature dependence

$$\frac{I_{SN}}{I_{NN}} = 2 \sum_{m=1}^{\infty} (-1)^{(m+1)} \frac{m\Delta_1}{\kappa_B T} K_1 \frac{(m\Delta_1)}{\kappa_B T} ,$$

where Δ_1 is the superconducting energy gap and K_1 is a modified Bessel function of the second kind. In a recent experiment, Bakker, van Kemper, and Wyder (15) described an alternative procedure in which they fit the measured I-V curve to temperature-dependent computer-generated curves for $Al-Al_2O_3-Ag$ junctions in the region 0.1 - 1.0 K. They found that the temperatures so measured agreed within a few mK with calibrated germanium thermometers.

NOISE THERMOMETRY

In another thermometry scheme, suggested by Kamper (16), the thermal noise spectrum in a resistance is detected by use of a superconducting Josephson Junction. The principle of the noise thermometer is that the voltage drop across a current-carrying resistor fluctuates in time due to random thermal excitations or "Johnson noise" within it, thus producing a temperature-dependent effect. The equation for the Johnson noise voltage is as follows:

$$< V^2 > = 4\kappa_B TRB ,$$

where B is the detector bandwidth. Now, a Josephson junction has the property that it radiates a frequency characterized by the relation

$$\nu = \frac{2eV}{hc} (\text{independent of temperature, } T < T_c)$$

Thus, if the time-varying voltage from the resistance is applied to the Josephson junction, one can detect a time-varying frequency in its radiation. By sampling the spectrum of the junction radiation with a frequency counter, one can determine the temperature from the mean square deviation, σ^2, of the frequency, by using the relation

$$T = \sigma^2 \tau \varphi_0^2 / 2\kappa_B R,$$

where τ is the gate time of the detector and φ_0 is the flux quantum.

This derivation depends upon the impedance of the junction being much larger than that of the resistor, and upon the absence of interfering sources of noise, but it has the considerable virtue that the temperature is calculable directly; it is, potentially at least, an absolute thermometer.

In experiments at NBS Boulder, the feasibility of the scheme was established by Kamper, Radebaugh, Siegwarth, and Zimmerman (17). In further experiments at NBS Washington, Soulen (17) has found agreement to within a few millikelvin in the range 20-200 mK between a noise thermometer and a cerous magnesium nitrate paramagnetic salt thermometer which was calibrated against the ^{3}He vapor-pressure scale.

In collaboration with H. Marshak, Soulen has compared the same noise thermometer with a ^{60}Co gamma-ray anisotropy thermometer. Between 20 mK and 50 mK, the two measurements of temperature agreed to within one millikelvin (17).

REFERENCES

1. W. S. Corak, B. B. Goodman, C. B. Satterthwaite, and A. Wexler, Phys. Rev. 102, 656 (1956).

2. C. J. Gorter and H. B. G. Casimir, Z. Physik. 35, 963 (1934).

3. H. Frohlich, Proc. Roy. Soc. (London) A223, 296 (1954), C. G. Kuper, ibid, A227, 214 (1955), and J. Bardeen, Phys. Rev. 97, 1724 (1955).

4. H. van Dijk and D. Schoenberg, Nature 164, 151 (1949). See also F. G. Brickwedde, H. van Dijk, M. Durieux, J. R. Clement, and J. K. Logan, J. Res. NBS 64A, 1 (1960).

5. J. Bardeen, L. N. Cooper, and J. R. Schrieffer, Phys. Rev. 108, 1175 (1957).

6. The early progress of the critical field-temperature relation can be traced through the paper of J. C. Swihart, D. J. Scalapino, and Y. Wada, Phys. Rev. Lett. 14, 106, (1965).

7. E. P. Harris and D. E. Mapother, Phys. Rev., 165, 522 (1968).

8. R. P. Ries and D. E. Mapother, paper O-3, Fifth Symposium on Temperature, NBS Washington, D. C., 21-24 June, 1971.

9. International Committee on Weights and Measures, Metrologia, 5, 35 (1969).

10. W. D. Gregory, Phys. Rev. 165, 556 (1968).

11. J. F. Schooley and R. J. Soulen, Jr., paper S-12, Fifth Symposium on Temperature, NBS Washington, D. C., 21-24 June, 1971.

12. R. J. Soulen, Jr., and J. H. Colwell, J. Low Temp. Phys. (to be published).

13. I. Giaver and K. Megerle, Phys. Rev. 122, 1101 (1961).

14. D. H. Douglass, Jr., and L. M. Falicov, Prog. Low Temp. Phys. IV, p 143, (1964), ed by C. J. Gorter.

15. J. W. Bakker, H. van Kempen, and P. Wyder, paper O-4, Fifth Symposium on Temperature, NBS Washington, D. C., 21-24 June, 1971.

16. R. A. Kamper, paper M1 in Symposium on the Physics of Superconducting Devices, Univ. of Virginia, 28-29 Apr. 1967. See also R. A. Kamper, and J. E. Zimmerman, J. Appl. Phys. 42, 132 (1971).

17. Private communication.

MILLIMETER AND SUBMILLIMETER DETECTORS AND DEVICES

Sidney Shapiro

Department of Electrical Engineering

University of Rochester

I. INTRODUCTION

The non-linear properties of the ac Josephson effect[1] have been demonstrated in the laboratory by a number of experiments in which the interaction of a Josephson junction with external radiation, or with a resonant cavity, or with both, has been the focus of attention. Experiments have been reported in which Josephson junctions were employed in harmonic generation,[2][3][4] mixing,[5][6][7] generation and self-mixing,[8] and far-infrared detection.[9][10] As a detector, there have been quoted sensitivities of 5×15^{-15} Watt,[10] speeds of better than 10 nanosecond,[9] and frequency response extending from very long wavelengths to better than 100 microns.[9] As a mixer, numbers that have been reported include conversion loss of about 6 dB at 13 GHz,[11] detection of a signal at 891 GHz by beating it against the 84th harmonic of a signal at 10.6 GHz,[6] and upconversion from 20 GHz to about 500 GHz.[8] None of these figures appears to represent a device limit. Other activity is also underway. The ac Josephson effect is the heart of a system to maintain the legal volt in terms of frequency and hence improve the reliability of standards intercomparisons around the world.[12] Several laboratories are at work on development of even more sensitive Josephson effect detectors for application in areas such as far-infrared astronomy.[13]

In this lecture we shall review some of the basic principles underlying these applications of the ac Josephson effect. In particular we shall try to show why the Josephson effect can be used as a far-infrared detector and mixer, how it has been used up to now in the laboratory, and what some of the inherent and practical limitations are to attainable performance. In carrying out this program, some topics will have to be ignored because of lack of time, including consideration of noise in Josephson junctions.[14]

Our approach will involve first the study of an ideal Josephson junction. By definition the current flowing through an ideal junction is entirely described by Josephson's phenomenological equations. Without this current we have no Josephson effect and that is reason enough for beginning our study here. Yet a real junction is more complex. We shall discuss how we must modify our concept of the ideal junction in order to understand the electrical characteristics of real junctions and shall touch briefly on the affect that measuring circuitry has on the observed characteristics of real junctions.

Throughout this development, very little will be said about the microscopic theory of the Josephson effect.[15][16] The phenomenological description will be found to be entirely adequate to understand the high frequency behavior of Josephson junctions with the sole exception of the Riedel peak phenomenon[17] - i.e., the peak in the amplitude of the ac Josephson current that occurs at a bias voltage corresponding to the superconducting energy gap for the superconductor forming the junction. An ad hoc modification of the phenomenological equations is sufficient to incorporate the Riedel peak phenomenon.[18][19] The question of the inherent frequency response of Josephson junctions is treated on this basis.

We conclude with a commentary on the directions in which future work is desirable.

II. PHENOMENOLOGICAL DESCRIPTION OF THE AC JOSEPHSON EFFECT

A. Junction with Incident Radiation

A variety of configurations have been used to study the Josephson effect, all of which we shall refer to as Josephson junctions. Some of the possible junction configurations are shown in Fig. 1. The point contact, shown in Fig. 1D, has been used for much of the high frequency work. It is readily formed by lightly bringing together a flat of one superconductor and a point of the same or another superconductor. Frequently niobium is used; but lead or tin or solder serve well too. In the point contact the ingredient that yields the Josephson effects is the constriction. It represents a barrier to the flow of current between the two pieces of superconductor. In the Josephson phenomena one is always concerned with the passage of current from one essentially

MILLIMETER AND SUBMILLIMETER DETECTORS AND DEVICES

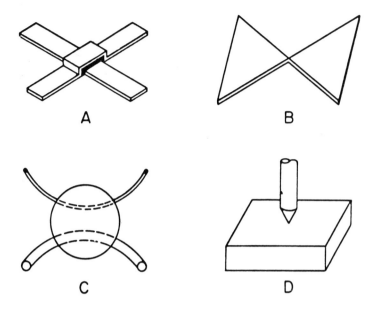

Fig. 1. Several different forms of Josephson junction.

bulk piece of superconductor to another essentially bulk piece of superconductor through some kind of a barrier. In the point contact it is the geometrical constriction. In each of the other configurations there is a corresponding barrier. In the thin-film form, for instance, Fig. 1A, the two superconducting films are separated by a real dielectric barrier consisting of 10 or 20 Angstroms or so of material. Yet current can flow from one film to the other through the barrier.

In the ideal junction the current that flows through the barrier is entirely Josephson current. It is important to appreciate how the Josephson current depends on voltage and on time. Although other parameters such as magnetic field and temperature also enter, the voltage and time dependences dominate the ac Josephson effect and we shall concentrate on them.

In the detailed microscopic quantum-mechanical theory the wavefunctions for the superconductors on each side of the barrier enter as do matrix elements coupling them across the barrier. Josephson has shown,[1] however, that the theory can be drastically simplified and put into parametric form provided the coupling across the barrier is weak. What this means is that the Josephson current should not be too large - say, below one milliampere or so.

We write then that the Josephson current, I_J, depends on a parameter, ϕ, i.e., $I_J = I_J(\phi)$, and ϕ in turn depends on voltage and time, i.e., $\phi = \phi(V,t)$. The parameter ϕ is in fact the difference in phase of the superconducting wavefunctions across the barrier but that is not important for what follows. Josephson showed that

$$I_J = I_o \sin[\phi(V,t)] \qquad (1)$$

and

$$\frac{d\phi}{dt} = \frac{2e}{\hbar} V \quad . \qquad (2)$$

When $V = 0$, ϕ is constant and we see that I_o is the maximum value of the zero-voltage Josephson current. We treat I_o as an experimentally determined number. Current can be driven through the junction from an external circuit and the junction will remain at zero-voltage provided this maximum value is not exceeded. If it is exceeded then voltage will be developed across the junction.

If the voltage is some constant dc value, $V = V_{DC}$, then ϕ varies linearly with time, $\phi = (2e/\hbar)V_{DC}t + \phi_o$, where the integration constant ϕ_o represents the initial phase difference across the junction. The current varies sinusoidally in time

$$I_J = I_o \sin(\omega_J t + \phi_o)$$

with $\quad \omega_J = (2e/\hbar)V_{DC} \quad .$ \qquad (3)

This is the ac Josephson effect where a dc voltage across the junction causes an alternating Josephson current to flow. The frequency, ω_J, at which this alternating current flows is proportional to the dc voltage with one millivolt yielding a frequency of about 484 GHz.

In order to see how this effect can be used to detect radiation, consider what happens when a monochromatic incident signal falls on the junction. The total voltage then will be the dc bias plus the rf signal at angular frequency ω

$$V = V_{DC} + V_{rf} \cos(\omega t) \quad . \qquad (4)$$

It follows that

$$\phi = \omega_J t + \left(\frac{2e}{\hbar}\right)\frac{V_{rf}}{\omega}\sin(\omega t) + \phi_o . \tag{5}$$

This in turn leads to a frequency modulation of the Josephson current as is apparent when Eq. 5 is inserted in Eq. 1 and the resulting expression is expanded using standard trigonometric identities. The final result is

$$I/I_o = \sum_n J_n(2eV_{rf}/\hbar\omega)\sin[(\omega_J + n\omega)t + \phi_o] \tag{6}$$

where J_n is the Bessel function of order n and the integer n ranges over all values.

In the absence of the applied rf, the Josephson current oscillates at the frequency ω_J. With the rf present there are additional frequency components introduced into the Josephson current; namely, side-bands at $\omega_J \pm |n|\omega$. Since ω_J is proportional to the bias voltage, one of the side-bands will occur at zero frequency - at dc - whenever

$$V_{DC} = \frac{n\hbar\omega}{2e}, \quad n = 0, \pm 1, \pm 2, \ldots \tag{7}$$

This means there will be a change in the Josephson current in the presence of rf and that change can be used to detect the rf. At those values of junction bias given by Eq. 7, the Josephson current will have a maximum dc component given by

$$I_{DC}/I_o = \sum_n |J_n(2eV_{rf}/\hbar\omega)|\delta(V_{DC} \pm n\hbar\omega/2e) \tag{8}$$

where $\delta(x) = 0$, $x \neq 0$ and $\delta(x) = 1$, $x = 0$.

A typical experimental result[5] corresponding to this situation is shown in Fig. 2 for a Nb-Nb point-contact junction exposed to various levels of 72 GHz signal. With no applied rf, the curve labelled "MAX", the zero-voltage Josephson current is visible as is, once the maximum amount is exceeded, the transition to the finite voltage part of the V-I curve. As increasing amounts of rf power are applied, steps in current appear in the V-I curve at those values of dc bias expected from Eq. 7. By looking across the figure for a particular step - a particular value of n - it is evident that the size of the current step depends on the rf voltage, whereas its location depends only on the applied frequency.

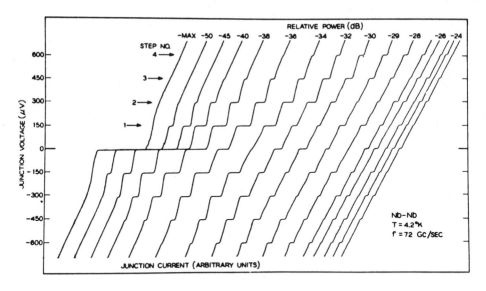

Fig. 2. V-I curves for a Nb-Nb point-contact junction showing rf-induced current steps.

Although not shown here the rf voltage dependence of the current in the steps of Fig. 2 are in reasonable agreement with the Bessel function variation expected from Eq. 8. (See Fig. 2 in reference 5.) Similar approximate agreement has also been obtained in experiments employing thin-film tunnel junctions.[20] Equation 7, on the other hand, has been shown to be exact to better than one part per million for different forms of junction, junctions employing different superconductors, for different step orders n spanning a wide interval, and for different applied frequencies from 10 GHz to nearly 1000 GHz.[12] It is these rf induced Josephson steps obeying so exactly the frequency-voltage proportionality of Eq. 7 that provide the basis for the voltage standard intercomparison system referred to above.

To see how the same effect has been used as a far-infrared detector for weak broad-band signals consider Eq. 8 once again and focus on the step at zero-voltage, $V_{DC} = 0$, n = 0. For low powers the maximum zero-voltage current is decreased initially by applied radiation. Broad-band radiation then will not induce additional current steps but it will change the maximum zero-voltage current. To use this effect as a detector it is necessary to measure with high sensitivity <u>not</u> a change in current but rather a change in <u>maximum</u> zero-voltage current.

Experiments of this type have been done and, among other things, have provided useful experimental information on the frequency response of Josephson junctions. Details of the experimental techniques are thoroughly described in the references[9][10] and will not be repeated here. The key result of the broad-band detector experiments is that a Josephson junction responds to applied radiation at frequencies much higher than that corresponding to the superconducting energy gap of the superconductor of which it is made - a result confirmed also by experiments using single frequency far-infrared laser sources.[4] We shall return to these results below when we discuss the Riedel peak phenomenon.

We turn now to the use of the ac Josephson effect as a mixer of two high frequency signals at angular frequencies ω_1 and ω_2. The total voltage is now

$$V = V_{DC} + V_1 \cos(\omega_1 t + \theta_1) + V_2 \cos(\omega_2 t + \theta_2) \quad . \tag{9}$$

Using Eq. 9 in Eqs. 1 and 2 it follows that

$$I/I_o = \sum_k \sum_\ell J_k(2eV_1/\hbar\omega_1) J_\ell(2eV_2/\hbar\omega_2) \sin[(\omega_J + k\omega_1 + \ell\omega_2)t + \phi_o + k\theta_1 + \ell\theta_2] \tag{10}$$

where the integers k and ℓ range over all values.

Equation 10 is the general expression for the Josephson current in an ideal junction at all frequencies and bias voltages in the presence of two applied rf voltages. Consider the component of current at zero-frequency: in addition to the rf induced steps expected from each applied rf separately, steps associated with the difference frequency, $\Delta\omega = \omega_1 - \omega_2$, are also expected. These additional steps lie at values of ω_J, i.e., bias voltage, spaced by $\Delta\omega$ from each of the steps associated directly with the two input frequencies. Such steps have been observed experimentally[5] as shown for a point-contact junction in Fig. 3. Again the amplitude of a current step depends on the rf voltages whereas its location depends only on the applied frequencies. As is evident from Eq. 10, the maximum Josephson current in each step is now given by a product of Bessel functions. Reasonable agreement with the expected variation was found in the experiments. (See Fig. 5 of reference 5.)

Equation 10 also shows that the Josephson current has a component oscillating at the difference frequency $\Delta\omega$. It is the presence of this current that provides the basis for the use of the

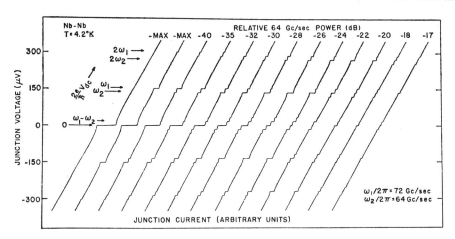

Fig. 3. The curve on the left is a V-I curve in the absence of microwave power. The remaining curves are a series of V-I curves with variable power from a source at 64 Gc/sec mixing with fixed power from a source at 72 Gc/sec. The constant-voltage steps are labelled with the Josephson angular frequency corresponding to the voltage at the step. The 72-Gc/sec power produces the steps at ω_1 and $2\omega_1$ while the 64-Gc/sec power produces the steps at ω_2 and $2\omega_2$. Mixing action produces the remaining steps. Note especially the additional steps at the difference frequency $(\omega_1-\omega_2)$ and at multiples of the difference frequency.

ac Josephson effect as a heterodyne detector of high frequency signals.

B. Junction with Resonant Cavity

We have been concerned so far with the effect on Josephson current of several forms of applied radiation. We now discuss the effect of the interaction of an ideal junction and a cavity resonant at a single high frequency. Equations 1 and 2 must now be supplemented by an equation which describes the simultaneous effect of the Josephson current exciting the cavity resonant mode electromagnetic fields and those fields in turn feeding back onto the junction voltage so as to modify the Josephson current.

Werthamer[16][21] has given an extensive discussion of this problem and has shown that for most purposes it is sufficient to treat the cavity as an harmonic oscillator driven by the Josephson current in the junction. The result is an equation for the junction voltage

$$\left(\frac{d^2}{dt^2} + \gamma \frac{d}{dt} + \omega_c^2\right) \int^t \Omega(t')dt' = \omega_p^2 \sin[\omega_J t + \phi_o + \int^t \Omega(t')dt'] \quad (11)$$

where the total junction voltage is

$$V = \hbar\Omega(t)/2e + V_{DC} \quad . \quad (12)$$

The cavity is described by its resonant frequency ω_c and its quality factor Q. The cavity damping constant γ is the ratio

$$\gamma = \omega_c/Q \quad . \quad (13)$$

We treat ω_p^2 as an empirical parameter that represents the degree of coupling between the junction current and the cavity fields.

The solutions of Eqs. 11 and 12 for various values of ω_c, Q, and ω_p^2 have been studied on an analog computer.[21] As might be expected, the only rf fields that have appreciable amplitude in the cavity are those near ω_c, provided the Q is greater than about ten - a condition readily satisfied in most experiments with cavities. Thus to a good approximation the presence of the cavity affects the junction only when there is a component of Josephson current at or near ω_c. Such a component certainly exists when

$$\omega_J \simeq \omega_c$$
$$\quad (14)$$
i.e., $\quad V_{DC} \simeq \hbar\omega_c/2e \quad .$

At bias voltages satisfying Eq. 14, the effect of the cavity can be determined by writing

$$\int^t \Omega(t')dt' = Z \sin(\omega_J t + \theta) \quad (15)$$

and solving Eqs. 11 and 15 self-consistently for the amplitude Z and phase θ. Without actually carrying through that procedure (see, however, references 10 and 21) we can readily see what to expect in the ideal junction dc voltage-current characteristic. As the bias increases through the value $V_{DC} = \hbar\omega_c/2e$, the dc Josephson current should increase from zero to a maximum value at $V_{DC} = \hbar\omega_c/2e$ and then fall again to zero. Experimentally this current "line" is usually observed as a current "step" with a spread in voltage related to the cavity Q. Thus the effect of a resonant cavity is to induce a somewhat smeared out current step in the junction characteristic analogous to the sharp rf-induced current steps.

When the junction is biased on the cavity-induced step, Eq. 14, the Josephson effect functions as a high frequency power generator. Real power from the bias circuit is converted to the cavity frequency and dissipated in the cavity losses. By coupling the cavity to an output transmission line in a conventional manner some of this power is available for driving an external load. Dayem and Grimes[22] used a point-contact junction in a coaxial cavity resonant at 9 GHz to produce about 10^{-10} Watt with a power conversion efficiency of about 0.1%. Both figures represented substantial improvement over earlier work in which thin-film tunnel junctions were used both for their Josephson effect and because under certain conditions they also serve as electromagnetic resonances.[1] [We note that the objective of this earlier work was to demonstrate for the first time that radiation was produced by the ac Josephson effect.]

C. Junction with Cavity and Incident Radiation

Although the numbers for power delivered to an external load are small, the current through and voltage across the junction at the cavity frequency are not small. This fact suggested the use of the Josephson effect in a novel way - called self-mixing - which has now been demonstrated. In this mode of operation the junction is still coupled to a resonant cavity but simultaneously an incident signal falls on the junction.

Richards and Sterling[10] studied the case where the frequency of the incident signal lies within the resonant bandwidth of the cavity. Their analysis showed the response of the junction to be narrowed considerably in bandwidth and enhanced considerably in sensitivity by the presence of the cavity by what amounts to a regenerative interaction. The detector that they constructed using this approach yielded the figure of 5×10^{-15} Watt NEP quoted above.

In contrast, the self-mixing mode[8] is concerned with a monochromatic incident signal whose frequency lies well outside the cavity bandwidth. By a slight extension of the discussion of Section II.B. we can deduce what to expect for this case in the ideal junction dc voltage-current characteristic. Rf-induced steps associated with the incident signal alone will be present at bias voltages given by Eq. 8 as will the cavity-induced step at the bias given by Eq. 14. The latter step arose because at that bias there existed a component of Josephson current at ω_c. In the presence of the incident signal at ω, there are two additional values of bias at which there exists a Josephson current component at ω_c; namely

$$\omega_J - \omega \simeq \omega_c \tag{16}$$

and $\quad\quad \omega - \omega_J \simeq \omega_c$.

At these bias values the Josephson frequency ω_J is, respectively, the sum and difference of the signal and cavity frequencies. Expressing this result explicitly in terms of bias voltage,

$$V_{DC} \simeq (\hbar/2e)(\omega \pm \omega_c) . \tag{17}$$

It is clear why this mode is called self-mixing: at the two values of bias given by Eq. 17, the Josephson junction plays the role of a local oscillator to mix with the incident signal and produce a difference frequency output at the cavity frequency. In essence, the Josephson effect allows heterodyne frequency conversion with just a battery and a cavity.

The line shape for these sum and difference steps can be obtained for an ideal junction by following the prescription presented in Section II.B. Again for real junctions a rounded current step is observed rather than a line as is apparent from Fig. 4 which was obtained for a point-contact junction mounted in a coaxial cavity resonant at about 20 GHz and driven by an incident signal at about 75 GHz.

Although the above discussion implies that the signal frequency is higher than the output, i.e., cavity, frequency (down-conversion), this is not required. Sum and difference steps have also been observed[8] in the V-I curve of a point-contact junction coupled to a stub resonant at about 500 GHz when 25 GHz radiation was incident (upconversion).

When the junction is biased on the sum or difference step, Eq. 17, the self-mixing effect converts real power from the incident signal to the cavity frequency. By coupling the cavity to an output transmission line some of this power is made available for driving an external load. Power levels of about 10^{-10} Watt have

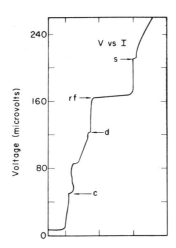

Fig. 4. V-I curve for a point-contact junction in a cavity resonant at 20 GHz and simultaneously driven by applied rf at 75 GHz. Constant-voltage steps are visible corresponding to cavity ("c") and applied rf ("rf") frequencies, and to their sum ("s") and difference ("d") frequencies.

been measured at 9.4 GHz, the cavity frequency, converted from an incident signal at 80 GHz.[23]

III. REAL JOSEPHSON JUNCTIONS

A. Introduction

In Section II attention was focussed on Josephson's phenomenological equations so as to identify the basic mechanisms at work in the various high frequency applications of the Josephson effect. That these equations alone are not adequate to account for the V-I curves of real junctions is apparent from the following considerations: The current contributions to the dc V-I curve (e.g., Eq. 8) occur either as sharp steps or as narrow lines at or near certain discrete values of bias voltage. At other values of bias the ideal junction of Section II carries no direct current. Yet any real junction carries direct current at all non-zero values of bias as pictured, e.g., in Figs. 2, 3 and 4. Thus to obtain a more realistic model of a real junction something must be added to the ideal junction that will result in current at these general bias values. The simplest possibility is to add a resistor shunting

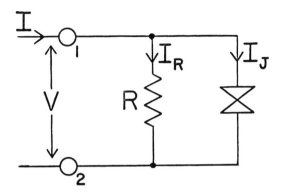

Fig. 5. Equivalent circuit model of a point-contact junction.

the ideal junction as shown in Fig. 5. Stewart,[24] and others,[25][26][27] have shown that the equivalent circuit model of Fig. 5 is a very satisfactory one for a point-contact junction, and we shall confine our attention to it. The more general case which includes a shunt capacitor is treated by the referenced authors.

B. Model of a Real Point-Contact Junction

We emphasize that the resistor R in Fig. 5 is inherent in the junction and represents internal current-carrying processes other than those described by Eqs. 1 and 2. For typical point-contact junctions the value of R falls in the range 0.01 ohm to 1 ohm. Two channels for current flow exist then: one is the Josephson channel represented by the ideal junction, the other the resistive channel represented by R. From a circuit element point of view, the ideal junction is a non-linear inductor, L_J, which is readily demonstrated by evaluating dI_J/dt:

$$dI_J/dt = I_o \cos\phi \, d\phi/dt$$
$$= (2e/\hbar)(I_o \cos\phi) V \tag{18}$$

or,
$$L_J = (\hbar/2eI_o)\sec\phi$$
$$= (\hbar/2eI_o)\sec \int^t V dt' \quad . \tag{19}$$

Returning to Fig. 5, we have

$$I = I_R + I_J = V/R + I_J \qquad (20)$$

$$I = (\hbar/2eR)d\phi/dt + I_o \sin\phi . \qquad (21)$$

We now suppose that the current into terminal 1 is only dc

$$I \equiv I_{DC} . \qquad (22)$$

This is a very accurate representation of typical experimental situations in which the bias supply at room temperature is connected to the junction in the liquid helium bath by a pair of long fine wires. The inductance of the wires decouples rf currents and voltages present in the junction from the bias supply. It is to model this real world situation that we set $I \equiv I_{DC}$. We note that this ansatz remains accurate even in the presence of applied radiation since the typical waveguide or cavity connection involves unavoidable - and desirable - rf bypass capacitance. We now suppose that the bias is provided from a source of high internal impedance relative to the internal junction resistance R - a situation easily achieved in practice in view of the rather low value of R for typical point-contact junctions - so that we may take I_{DC} as constant when Eq. 22 is substituted in Eq. 21. The value of I_{DC} will be set then by the external bias circuit. On rearranging Eq. 21 we have

$$d\phi/dt = (2e/\hbar) R[I_{DC} - I_o \sin\phi] . \qquad (23)$$

We see immediately from Eq. 23 that the solution $d\phi/dt = 0$, i.e., $V = 0$, is obtained provided $I_{DC} \leq I_o$. For this regime,

$$\begin{aligned} I_{DC} &= I_J = I_o \sin\phi_o . \\ I_R &= 0 \\ I_{DC} &\leq I_o \end{aligned} \qquad (24)$$

and the initial phase difference, ϕ_o, is determined by the value of I_{DC} set by the bias supply. The zero-voltage current is determined entirely by the Josephson channel.

The effect of the resistor is evident for $I_{DC} > I_o$. To display this we first solve Eq. 23 by direct integration using the integral formula

$$\int \frac{d\phi}{a+b\sin\phi} = \frac{1}{(a^2-b^2)^{1/2}} \sin^{-1}\left(\frac{b+a\sin\phi}{a+b\sin\phi}\right). \quad (25)$$

After some manipulation the result is conveniently written

$$\sin\phi = \frac{(I_{dc}/I_o)\sin\gamma + 1}{(I_{dc}/I_o) + \sin\gamma} \quad (26)$$

where

$$\gamma = \left(\frac{2e}{\hbar}\right) I_o R[(I_{dc}/I_o)^2 - 1]^{1/2} t + \alpha \quad (27)$$

and α is an integration constant.

By substituting Eqs. 26 and 27 in Eq. 23, the expression for the full time-dependent voltage across the junction may be displayed. We are primarily interested here, however, in the dc V-I curve which requires a time-average of the resulting expression. The mathematics is straightforward though tedious and will not be presented here. The final result displays the relationship between dc voltage and dc current as the equation of an hyperbola,

$$\left(\frac{I_{DC}}{I_o}\right)^2 - \left(\frac{V_{DC}}{I_o R}\right)^2 = 1. \quad (28)$$

Equation 28 is plotted in Fig. 6(a). The marked effect of the resistive channel is evident for non-zero voltages: the ideal junction connected only to an external bias supply carries direct current only at zero voltage, the real junction carries direct current at all voltages. At voltages greater than $I_o R$, nearly all the direct current is carried by the resistive channel. Hence the current in the Josephson channel, I_J, is virtually sinusoidal as required by our ideal junction analysis.

At voltages less than $I_o R$, however, and especially at voltages only slightly above zero, the amount of direct current that can be carried by the resistive channel is small. (See the dashed line in Fig. 6(a).) I_J is forced to have a large dc component. Hence its waveform in time differs radically from sinusoidal and is rich in harmonics.[26] Consequently it is in this region of bias that we would expect strong interactions of the junction with applied radiation and resonant cavities as is indeed observed to be the case. In order then to have useful Josephson effects at higher

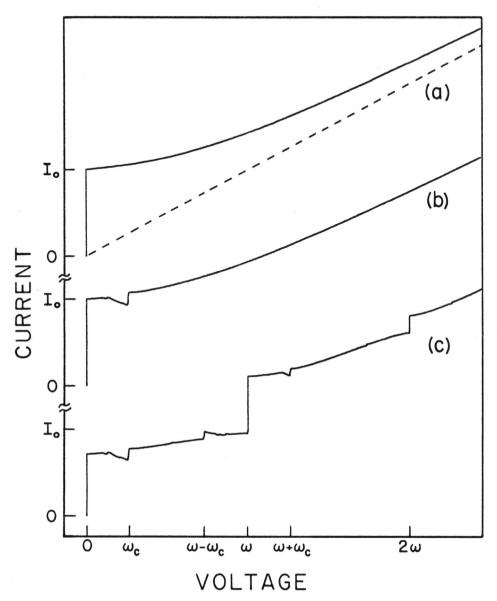

Fig. 6. I-V curves for (a) the junction model of Fig. 5; (b) with coupling to resonant cavity added; and (c) with external rf added.

and higher frequencies it is necessary to expand the voltage range of this bias regime and so we must seek junctions with a large value of the product $I_o R$.

Of course at high frequencies the effect of junction capacitance cannot be ignored. The more complete analyses[24]-[27] demonstrate that junction capacitance leads to hysteresis in the dc V-I curve - an effect that hampers practical use of such junctions. To avoid such hysteresis the condition $I_o R \ll \hbar/2eRC$ must also be satisfied. Since $\hbar/2e$ is about 2×10^{-15} Volts/Hz, for a junction with I_o of about 0.3 mA and R of about 1 ohm - and hence $I_o R$ corresponding to about 300 GHz or 1 mm - we desire a C of less than 0.01 picofarad.

C. Interaction with Radiation and Resonant Cavities

Although we shall not go into any of the details here, models have been considered in which to the junction of Fig. 5 is added both coupling to an external resonator[28][29] and drive from an external rf source.[28][29][30][31] The criterion is always to obtain V-I curves from the model corresponding as closely as possible to those actually observed in experiments. Figure 6 shows a set of curves obtained from one such model[29] in which the junction of Fig. 5 and Eq. 28 was inductively coupled to a parallel R-L-C resonance at ω_c and driven from a nearly constant current rf source at frequency ω.

The resulting curves, plotted by the analog computer used to solve the circuit equations, correspond almost exactly to experimental curves we have obtained in our laboratory. They show the modifications obtained for real point-contact junctions in the V-I curves expected on the basis of our earlier discussion of ideal junction characteristics.

By use of a spectrum analyzer we have also observed radiation out of the cavity when the junction was biased to the cavity or sum or difference step as expected from our earlier discussion. The radiation bandwidth is very narrow; much narrower than the cavity bandwidth. This is connected with the existence of the steps in the V-I curve since fluctuations in bias current lead to very small fluctuations in junction voltage on the step and hence in output frequency. The performance of the regenerative detector and of the self-mixer both benefit from this effect.

D. Frequency Dependence of the Josephson Current

The phenomenological equations and equivalent circuits developed to this point ignore one effect that is predicted by microscopic theory[16][17] and that has been confirmed experimentally;[18][19][32] namely, the frequency dependence of the Josephson current, or Riedel peak phenomenon.

It is well known that as frequency increases into the optical region differences between the normal and superconducting states become unobservable. Similarly the Josephson effects should decrease in strength at high enough frequencies. This implies that the amplitude of the Josephson current, I_o in our previous equations, should itself be a function of frequency that tends to zero as frequency increases. This is in fact the case (see insert in Fig. 7) but before I_o drops it first rises to a peak, the Riedel peak, at a frequency corresponding to the Josephson frequency for a bias voltage of $2\Delta/e$, the superconducting energy gap. This is shown in Fig. 7 along with experimental data points obtained from studies of the effect of the peak on rf-induced step amplitudes.

The observed peak is very narrow and not especially high. In the present context its importance lies in showing that I_o remains appreciable relative to its zero-frequency value out to frequencies about twice that at the gap; namely, about $8\Delta/h$. For tin with a value of 2Δ of about 1 mV this corresponds to a frequency of about 1000 GHz or a wavelength of about 0.3 mm.

IV. SUMMARY AND COMMENTARY

In this lecture we have touched on only a few of the topics of importance for millimeter and submillimeter detectors and devices. We have demonstrated that there are a variety of ways in which to make use of the Josephson effect at high frequencies including broad-band and regenerative detection, conventional heterodyning,[33] and self-mixing. In discussing the latter we chose to emphasize the frequency conversion point-of-view. Alternatively a parametric amplifier approach is applicable.[34][35][36] Such studies as have been carried out along these lines suggest that there is more to be discovered about the basic Josephson processes and the nature of energy exchange with external circuits.

Certainly the Josephson phenomena have resisted for a long time characterization in the form of equivalent circuits. Much remains to be done along these lines especially as our understanding of these frequently subtle and certainly complex non-linear processes grows. Of particular importance will be analyses - and confirming experiments - that pinpoint optimum conditions for various modes of operation.

But the heart of any Josephson effect detector system is the junction. Although existing materials such as tin and niobium should be entirely satisfactory down to wavelengths of 1 mm and even 100 microns, other materials such as niobium nitride with larger energy gaps need to be used to reach to 10 microns. Yet

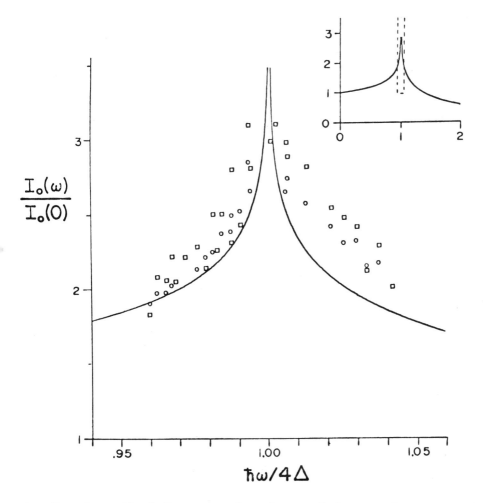

Fig. 7. Theoretical frequency dependence of the Josephson current (solid line) and data from experimental measurements of rf-induced step amplitudes. The dashed box in the insert shows the region plotted.

even more important than extending the frequency range is the need to learn how to tailor junctions to operate best under pre-determined conditions. Here the need to learn how to make junctions with specified values of I_o, or R, and even of junction capacitance is evident. Work along these lines is in its infancy. Perhaps in the course of such work solutions will be developed to the thorny problems of providing junctions that are stable, reliable, and reproducible.[37][38]

V. ACKNOWLEDGEMENT

It is a pleasure to acknowledge my indebtedness to close and stimulating interaction with many colleagues and students past and present. The long hours of give and take were invaluable to me in helping develop what understanding of the Josephson effect I possess.

VI. REFERENCES

The Josephson effect has been with us for a decade and its literature is immense in magnitude and in scope. Inevitably this list of references omits many significant contributions.

(1) Among the many review articles on the Josephson effect we note especially the following:

   (a) B.D. Josephson, Advances in Physics 14, 419 (1965).

   (b) P.W. Anderson in Progress in Low Temperature Physics Vol. V, edited by C.J. Gorter (North-Holland Publishing Co., Amsterdam, 1967), p. 1.

   (c) D.N. Langenberg, D.J. Scalapino, and B.N. Taylor, Proc. IEEE 54, 560 (1966).

   (d) P.L. Richards in Semiconductors and Semimetals, Vol. 6, Physics of III-V Compounds, edited by R.K. Willardson and A.C. Beer (Academic Press, N.Y., 1971).

(2) Sidney Shapiro, J. Appl. Phys. 38, 1879 (1967).

(3) G.R.S. Seraphim and R.C. McDermott, Phys. Letters 32A, 35 (1970).

(4) D.G. McDonald, et al., Appl. Phys. Letters 15, 121 (1969).

(5) C.C. Grimes and Sidney Shapiro, Phys. Rev. 169, 397 (1968).

(6) D.G. McDonald, et al., Appl. Phys. Letters 18, 162 (1971).

(7) A.J. DiNardo and E. Sard, J. Appl. Phys. 42, 105 (1971).

(8) Andrew Longacre, Jr., and Sidney Shapiro in Proceedings of the Symposium on Submillimeter Waves, Vol. XX of the Microwave Research Institute Symposia Series, edited by J. Fox (Polytechnic Press of the Polytechnic Institute of Brooklyn, N.Y., 1970) p. 295. See also, A.H. Silver and J.E. Zimmerman, Appl. Phys. Letters 10, 142 (1967).

(9) C.C. Grimes, P.L. Richards and Sidney Shapiro, J. Appl. Phys. 39, 3905 (1968).

(10) P.L. Richards and S.A. Sterling, Appl. Phys. Letters 14, 394 (1969). See also S.A. Sterling, Ph.D. Thesis, "The Interaction of High Frequency Electromagnetic Radiation of Superconducting Point Contact Junctions," report UCRL-19096 of Lawrence Radiation Laboratory, Univ. of California, Berkeley, under AEC contract No. W-7405-eng-48.

(11) Ref. 5, footnote 12.

(12) Extensive references to the literature on e/h measurements are contained in the review by J. Clarke, Am. J. Phys. 38, 1071 (1970).

(13) B. Ulrich, J. Appl. Phys. 42, 2 (1971).

(14) Among the many pertinent references we note only the following: M.J. Stephen, Phys. Rev. 186, 393 (1969); W.H. Henkels and W.W. Webb, Phys. Rev. Letters 26, 1164 (1971); A.H. Silver, J.E. Zimmerman, and R.A. Kamper, Appl. Phys. Letters 11, 209 (1967); and H. Kanter and F.L. Vernon, Jr., Phys. Rev. Letters 25, 588 (1970).

(15) B.D. Josephson, Phys. Letters 1, 251 (1962).

(16) N.R. Werthamer, Phys. Rev. 147, 255 (1966).

(17) E. von Riedel, Z. Naturforsch. 19A, 1634 (1964).

(18) C.A. Hamilton and Sidney Shapiro, Phys. Rev. Letters 26, 426 (1971).

(19) C.A. Hamilton, Phys. Rev. (to be published).

(20) C.A. Hamilton and Sidney Shapiro, Phys. Rev. B 2, 4494 (1970).

(21) N.R. Werthamer and Sidney Shapiro, Phys. Rev. 164, 523 (1967).

(22) A.H. Dayem and C.C. Grimes, Appl. Phys. Letters 9, 47 (1966). See also J.E. Zimmerman, J.A. Cowen, and A.H. Silver, Appl. Phys. Letters 9, 353 (1966).

(23) Andrew Longacre, Jr., and Sidney Shapiro, unpublished.

(24) W.D. Stewart, Appl. Phys. Letters 12, 277 (1968).

(25) D.E. McCumber, J. Appl. Phys. 39, 2503 (1968).

(26) D.B. Sullivan, et al., J. Appl. Phys. 41, 4865 (1970).

(27) L.G. Aslamazov and A.I. Larkin, Soviet Physics--JETP Letters 9, 87 (1969).

(28) C.A. Hamilton, Rev. Sci. Inst. (to be published).

(29) Andrew Longacre, Jr., and Sidney Shapiro (to be published).

(30) D.B. Sullivan and J.E. Zimmerman, J. Appl. Phys. (to be published).

(31) H. Fack and V. Kose, J. Appl. Phys. 42, 320 (1971); ibid., 42, 322 (1971).

(32) S.A. Buckner and T.F. Finnegan, Bull. Am. Phys. Soc. II, 16, 399 (1971).

(33) In addition to the earlier referenced works see also J.E. Zimmerman, J. Appl. Phys. 41, 1589 (1970).

(34) H. Zimmer, Appl. Phys. Letters 10, 193 (1967).

(35) P. Russer, Arch. Eleck. Übertragung 23, 417 (1969); Proc. IEEE 59, 282 (1971).

(36) A.N. Vystavkin, et al., Radioengineering and Electronic Physics 15, 2404 (1970); A.N. Vystavkin, et al., Proceedings of the Conference on High Frequency Generation and Amplification, Cornell Univ., August 1971 (to be published).

(37) W. Schroen, J. Appl. Phys. 39, 2671 (1968).

(38) J.E. Zimmerman, Paul Thiene, and J.T. Harding, J. Appl. Phys. 41, 1572 (1970).

MAGNETOMETERS AND INTERFERENCE DEVICES

Watt W. Webb

School of Applied and Engineering Physics, and Laboratory

for Atomic and Solid State Physics, Cornell University

I. INTRODUCTION

Superconducting quantum flux sensors presently provide the basis for instruments with capability for rapid measurement of magnetic fields as small as 10^{-10} Gauss; furthermore, electric potential differences as small as 10^{-16} volts, electrical currents, and resistances in low impedance circuits can be measured at power levels as low as 10^{-23} watts. These instruments are sufficiently well developed that commercial versions of the magnetometers are available from at least three commercial suppliers, and the cost of a complete magnetometer can be as low as a few thousand dollars. Many applications to physical measurement problems have been described in the literature and applications outside of physics in medicine and geophysics have been demonstrated. Superconducting quantum flux sensors provide new measurement capabilities that are ready for application to opportunities in many fields of science and technology.

A superconducting ring of small inductance (say 10^{-9}H) containing a weakly superconducting link that restricts the maximum or critical supercurrent to small values (say 10^{-5}A) is the basic element of these instruments. The long range quantum phase coherence characteristic of the superconducting state fixes quantized states of the ring. Coupling of the circulating supercurrent in the weak link and the magnetic flux in the ring through their part in the gauge invarient vector potential leads to modulation of the critical current in the weak link by the flux threading through the ring with a basic period of the flux quantum $\Phi_0 = 2\times10^{-15}$ Weber or 2×10^{-7} Gauss cm^2.

In this chapter the mechanisms and operation of one class of superconducting flux sensors is described in elementary terms using the basic properties of superconductors introduced in earlier chapters of this book. The operation of the most popular present device is outlined, and instrument configurations for measurements of magnetic fields, magnetic susceptibility, voltage and resistance are described. A brief discussion of noise problems and other limitations as presently understood completes the description of the instruments and the chapter closes with a discussion of some applications and possibilities for the future.

II. THEORY OF SUPERCONDUCTING QUANTUM FLUX SENSORS

The superconducting flux detector consists of three parts: first, a ring of good superconductor of small inductance (say L 10^{-9} H) closed by one or more "weak links" in which the maximum or critical supercurrent is restricted to some small value; second, a circuit to bias the current in the ring so that it sometimes exceeds the critical current; and third, an electronic circuit sensitive to the voltage appearing across the weak link whenever the bias current exceeds the critical supercurrent. It is essential that the critical current of the weakly superconducting ring be a quasiperiodic function of the magnetic flux threading through the ring. This section discusses this phenomenon in a necessarily elementary way; however several remarkably thorough descriptions have been given in the literature. Especially recommended is a long series of articles by A. H. Silver and J. E. Zimmerman and their coworkers who are responsible for most of the scientific advances in this field; their article with the title, "Quantum states in weakly connected superconducting rings",[1] is particularly germane. A review article by R. deBruyn Ouboter and A. Th. A. M. DeWaele[2] summarizing particularly the work at Leiden is also a very useful source for some of the subtle properties of quantum sensors. An earlier review of quantum superconductivity by P. W. Anderson is useful for its development of a physical viewpoint.[3]

The periodic behavior of a superconducting ring with a weak link arises from the long range phase coherence characterizing the superconducting state. It requires that the wave function $\psi = |\psi| e^{i\phi}$ describing the superconducting condensate be single valued. This condition can be satisfied by application of the Sommerfeld condition; that is, quantization of the canonical momentum

$$\overline{P}_s = 2m\overline{v}_s + 2e\overline{A} \qquad \text{II-1}$$

where m and e are the mass and charge of the electron so that

$$\oint_{ring} \overline{P} \cdot d\overline{s} = \oint 2m\overline{v}_s \cdot d\overline{s} + \oint 2e\overline{A} \cdot d\overline{s} = nh \qquad \text{II-2}$$

where $d\bar{s}$ is a line element of the ring, h is Planck's constant and n is an integer. The second integral can be evaluated by applying Stokes law to the line integral of the vector potential so that

$$\oint \bar{A} \cdot d\bar{s} = \int (\nabla \times \bar{A}) \cdot d\bar{a} = \int \bar{H} \cdot d\bar{a} = \Phi$$

is just the flux Φ threading the ring; that is, the integral of the magnetic field H over the area of the ring. The drift velocity in the first integral can be expressed in terms of the supercurrent $i_s = ev_s\sigma$ where n_s is the number density of superconducting electrons provided that σ the cross-sectional area is sufficiently small. If the ring is thick everywhere except at the weak link of length ℓ and cross-section σ, the only significant contribution to the velocity integral comes from the weak link since a path of integration along which $v_s = 0$ is available in the thick part. Thus $(1/e)\oint \bar{P} \cdot d\bar{s}$ becomes

$$\int (m/n_s e^2 \sigma) i_s \cdot ds = (m/n_s e^2)(\ell/\sigma) i_s$$

and Eq. II-2 can be written as

$$\left(\frac{m}{n_s e^2}\right)\left(\frac{\ell}{\sigma}\right) i_s + \Phi = n\frac{h}{2e} \qquad \text{II-3}$$

Defining the flux quantum as

$$\Phi_o \equiv \frac{h}{2e} = \begin{cases} 2 \times 10^{-7} \text{Gauss cm}^2 \\ 2 \times 10^{-15} \text{Weber} \end{cases} \qquad \text{II-4}$$

and the penetration depth λ as

$$\lambda^2 \equiv \frac{1}{4\pi} \frac{m}{e^2 n_s} \quad , \qquad \text{II-5}$$

we obtain the fluxoid quantization condition

$$4\pi\lambda^2 \left(\frac{\ell}{\sigma}\right) i_s + \Phi = n\Phi_o \qquad . \qquad \text{II-6}$$

Clearly the left hand side of this equation, called the fluxoid, has periodicity Φ_o. But the flux alone is exactly periodic only if $i_s = 0$, and only in that case is the <u>flux</u> quantized exactly.

For circuit analysis it's useful to identify the supercurrent contribution to the fluxoid as $L_s i_s$ using a "kinetic inductance", $L_s = 4\pi\lambda^2(\ell/\sigma)$, and to define the ratio of the kinetic inductance to the electromagnetic inductance as a dimensionless parameter

$$\gamma \equiv L_s/L \qquad . \qquad \text{II-7}$$

Although the penetration depth, kinetic inductance and γ all should depend on the current density i_s/σ and on the structure of the weak link, it is useful to neglect that dependence and treat γ as a current independent quantity to obtain a simpler linear approximation.

Assuming implicitly the appropriateness of a constant $\gamma = L_s/L$, Eq. II-6 can be written as

$$(L_s + L) i_s = (\gamma + 1) i_s = n\phi_o \quad . \qquad \text{II-8}$$

To describe an external applied field the appropriate variable is the flux ϕ_x due to the applied field that is intercepted by the ring. Thus the total flux consists of two contributions, the applied flux and the circulating current.

$$\phi = \phi_x + Li_s \qquad \text{II-9}$$

Thus with Eq. II-8, the current and flux are respectively

$$- Li_s = \frac{\phi_x - n\phi_o}{1 + \gamma} \qquad \text{II-10a}$$

and

$$\phi = \frac{\gamma \phi_x + n\phi_o}{1 + \gamma} \quad . \qquad \text{II-10b}$$

The second essential ingredient of the properties of the weakly superconducting ring is the critical current of the weak link. A second dimensionless parameter to specify the critical current in terms of flux excluded is also useful and the quantity $\beta \equiv 2(1 + \gamma) \cdot (Li_c/\phi_o)$ serves this purpose in the linearized case. If $\beta = 1$ and $\gamma = 0$, the critical current is just sufficient to exclude $\phi_o/2$ of applied flux.

The operation of the quantum flux detector depends on transitions between the states specified by Eqs. II-10. Fig. II-1, taken from Silver and Zimmerman[1] shows these solutions for three values of Li_c/ϕ_o with $\gamma = 10^{-3}(Li_c)^{-1}$. There are two classes of solutions. If the critical current is just enough to exclude half a flux quantum ($\beta = 1$) as in the center column of the figure, the current has a saw-tooth form as a function of applied flux and the flux in the ring is always an integer number of quanta. For $\beta < 1$ there are sections in the plot of I/ϕ_o where the fluxoid would not be quantized. For $\beta > 1$, as in the right hand column, the solutions are multiple valued and both current and flux follow hysteresis loops with transitions occuring at the termination of the allowed range of each quantum state as the applied flux is swept.

In the linear limit the Gibb's free energy of the system in

each state n varies parabolically as a function of applied flux around the minima at $\Phi_x = n\Phi_o$. For the state with n flux quanta, the free energy with respect to a zero conveniently taken at $i = i_c$ is

$$G = \frac{(\Phi_x - n\Phi_o)^2}{2L(1+\gamma)} - \frac{L(1+\gamma)}{2} i_c^2 \qquad \text{II-10c}$$

The first term is the energy associated with the circulating current needed to exclude the flux $I_x - n\Phi_o$ and the second term which is equal to $\beta i_c \Phi_o/4$ is the normalization constant making $G = 0$ at $i_s = i_c$. The bottom row of Fig. II-1 shows non-overlapping parabolas for $\beta \leq 1$, but for $\beta > 1$ the parabolas overlap. The transitions shown by the dotted lines describe the case where the system remains in a state until the critical current is exceeded in spite of the lower free energy of an adjacent state. This hysteretic behavior would be expected if the energy barrier between states is quite large -- a property which is observed experimentally.

However, current dependence of L_s and γ is expected; in particular, if the weak link satisfies the DC Josephson relation for a small area junction, $i_s = i_J \sin\theta$ where θ is the phase difference of the superconducting wave function across the weak link, one can define[4] the particular current dependent form for γ as γ_J

$$\gamma_J \equiv (\Phi_o/2\pi L i_s) \sin^{-1}(i_s/i_J) \qquad , \qquad \text{II-11a}$$

and for the kinetic inductance

$$L_{s,J} \equiv \frac{\Phi_o}{2\pi i_s} \sin^{-1}(i_s/i_J) \qquad . \qquad \text{II-11b}$$

The kinetic inductances defined here are "integral definitions" in terms of total flux in a ring. Effective differential inductances for the Josephson junction are $V/(di_s/dt) = (\Phi_o/2\pi i_J)[1-(i_s/i_J)^2]^{-\frac{1}{2}}$, and in a ring $-d\Phi_x/di_s = L + (\Phi_o/2\pi i_J) \sec[(2\pi/\Phi_o)(Li + \Phi_x - n\Phi_o)]$.

The more realistic small Josephson junction model for the weak link shows more clearly the connections amongst the quantum states of the ring, but the equations are more difficult to solve.[4] For this case with $i_c \to i_J$ and γ specified by Eq. II-11a, the operating equations are

$$Li = -Li_J \sin\left[(2\pi/\Phi_o)(Li + \Phi_x - n\Phi_o)\right] \qquad \text{II-12a}$$

$$\Phi = \Phi_x - Li_J \sin\left[(2\pi/\Phi_o)(\Phi - n\Phi_o)\right] \qquad . \qquad \text{II-12b}$$

Solutions for (a) $i_J = \Phi_o/4\pi L$, (b) $i_J = \Phi_o/2\pi L$ and (c) $i_J = \Phi/\pi L$ are shown in Fig. II-2, again following Silver and Zimmerman: For

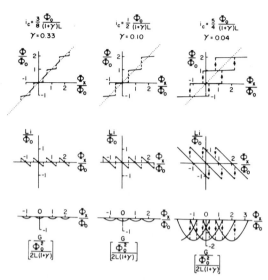

Fig. II-1. Normalized flux Φ/Φ_o, excluded flux Li/Φ_o and free energy $(G/\Phi_o^2)(2L(1+\gamma)$ calculated in the linearized model are given as functions of applied flux Φ_x/Φ_o for three levels of critical current i_c. After Silver and Zimmerman, Ref. 1.

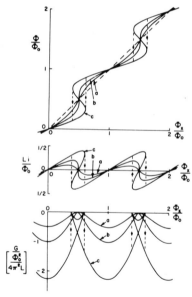

Fig. II-2. Normalized flux Φ/Φ_o, excluded flux Li/Φ_o and free energy $(G/\Phi_o^2)4\pi^2 L$ are given as functions of applied flux Φ_x/Φ_o for three Josephson critical current levels, a, b and c, where b denotes the case where one flux quantum can be excluded by the supercurrent, a less and c somewhat more. After Silver and Zimmerman, Ref. 1.

MAGNETOMETERS AND INTERFERENCE DEVICES   659

$i_c \lesssim \Phi_0/2\pi L$ only single valued solutions occur, and as in the linearized case for $\beta > 1$ the solutions are multiple valued, hysteresis is possible as shown by the dotted lines. The Gibb's free energy is given by the complicated equation II-12c below, but its behavior is quite similar to the linearized model discussed above.

$$G = -i_J \frac{\Phi_0}{2\pi} \cos \frac{2\pi}{\Phi_0}(\Phi-n\Phi_0) + \frac{1}{2}Li_J^2 \sin^2 \frac{2\pi}{\Phi_0}(\Phi-n\Phi_0) - G_c$$

where   II-12c

$$G_c = \begin{cases} -\dfrac{\Phi_0^2}{4\pi L} - \dfrac{1}{2}Li_J^2 & : \quad i_J \gtrsim \dfrac{\Phi_0}{2\pi L} \\ \dfrac{i_J \Phi_0}{2\pi} & : \quad i_J \lesssim \dfrac{\Phi_0}{2\pi L} \end{cases}$$

One physical feature obscured in the linearized case becomes clearer in the Josephson junction case -- that is the continuity of the functions describing the quantum states. If $i_J \lesssim \Phi/2\pi L$, the solutions of Eqs. II-11 clearly describe continuous behavior. In fact for $i_J < \Phi_0/2\pi$ increasing the applied flux simply produces a monatomic increase of flux in the ring which consists of a linear term with a small modulation superimposed on it by the current in the ring which varies between $-i_J \lesssim i_s \lesssim i_J$ to minimize the free energy, increasing or decreasing the flux as necessary. Clearly these are not distinct quantized <u>flux</u> states but the <u>fluxoid</u> is quantized throughout. What happens as $i_J \gg \Phi_0/2\pi L$, Φ/Φ_0 and Li/Φ_0 become increasingly skewed functions of Φ_x/Φ_0 describing a set of states of the superconducting ring in which the flux can be quantized in a set of equally spaced levels. For $i_J \simeq \Phi_0/2\pi L$ the magnitude of the change in flux between adjacent states at transition is not $\Phi_0/2$ because the kinetic energy contribution to the fluxoid integral is large at the transition.

Summary: Superconducting rings with a weak link show quantum-periodic response to an applied flux that may be continuous and reversible for low critical currents, or discontinuous and hysteretic for large critical currents. There is a range near $\beta \simeq 1$ where the selection rule $\Delta n = \pm 1$ applies and the transitions occur between adjacent quantum states of the fluxoid. It is this range of behavior which is used as a flux detector as outlined in the next section.

III. PRINCIPLES OF FLUX DETECTION USING A SUPER- CONDUCTING RING WITH A WEAK LINK

This section describes the operation of the flux detector that is presently most popular. Its essential features are outlined first: It consists basically of a superconducting ring coupled to a resonant L-C circuit excited by a radiofrequency oscillator near its resonant frequency. The RF oscillator level is adjusted until the maximum RF current induced in the superconducting ring exceeds the critical current (i_c or i_J) of the ring. Thus RF flux can enter the ring. The critical current is usually chosen $i_c \gtrsim \Phi_0/L$ so that a few flux quanta can be excluded.

The effective impedance of the RF resonant tank circuit is modified by the superconducting ring to which it is coupled. Since effective impedance of the ring depends on the total applied flux, the sum of RF and DC flux, the RF impedance of the tank circuit provides the desired measure of the DC flux applied to the ring. To facilitate the measurement of the DC flux it is modulated with an amplitude $\sim \frac{1}{2}\Phi_0$ at an audiofrequency and the detected RF voltage is then an audio frequency signal that is synchronously detected at this flux modulation frequency. Finally the detected signal is usually fed back to compensate for changes of the flux to be measured so that the instrument acts as a null detector in which the amplitude of the feed-back current provides a linear measure of the flux to be measured over a wide range of applied flux.

This terse description is amplified below. More detailed discussions are given in recent papers by Goodkind and Stolfa[5] and by Zimmerman, Thiene and Harding[6]. A few crucial details of the mechanisms of operation are still in doubt.

In this analysis several simplifying assumptions are used that may not be justified but which probably have little effect except on calculation of thermal fluctuations. For example, we shall assume that the response time $\tau \sim L/R_n$ of the superconducting ring is fast compared to the reciprocal oscillator frequency which is typically 1 to 30MHz. Typically $L \sim 10^{-9}$H and $R \sim 1\Omega$ so $\tau \lesssim 10^{-9}$ is usually expected under present operating conditions. Goodkind and Stolfa[5] have considered the alternative case of a long time constant.

The previous section described the response of the weakly superconducting ring in an applied field Φ_x. Next the effect of the ring on the response of the resonant tank circuit to which it is coupled must be considered. Only a qualitative discussion is given here. The effective circuit to be considered is given in Fig. III-1.

Here the superconducting loop of self inductance L_2 contains a weak link having a critical current function designated by the symbol λ, a kinetic inductance L_s and a shunt resistance R_n. The effective inductance L of the ring should take into account the effects of the coupling to the tank circuit through the mutual inductance M_{12} but this effect is usually neglected. Typical values of the parameters are $R_n \sim 1\Omega$, $L_2 \sim 10^{-9}$H. The shunt capacitance of the weak link is assumed to be negligibly small.

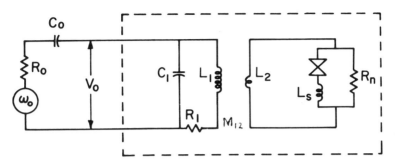

Fig. III-1. The effective circuit of a superconducting ring containing a weak link and coupled to a tank circuit is illustrated using notation defined in the text.

The tank circuit consisting of inductance L_1 with series resistance R_1 shunted by capacitance C_1 is driven at a frequency $\omega_1 \approx (LC)^{-\frac{1}{2}}$ by the oscillator with output impedance R_0 and C_0. The cold portion of the circuit is designated by surrounding dotted lines.

Two simple arguments give an idea of the mechanisms by which the tank circuit responds to the superconducting ring. With the ring perfectly superconducting it acts like a perfect diamagnet and it affects the impedance of the tank circuit through its inductive coupling only by slightly reducing the effective inductance of the tank circuit. This is a readily observable effect, but it is probably not the most appropriate picture for the operation of the flux sensor. A related mode of operation called the inductive mode by Goodkind and Stolfa uses the kinetic inductance of the weak link to modulate the tank circuit inductance.

In the most common mode of operation the effect of the weakly superconducting ring is supposed to be primarily an increase of dissipation and thus a lowering of the tank circuit Q. When the critical current is exceeded, dissipation begins in the weak link as its quantum states changes. Usually $\Phi_0/2L < i_c \lesssim 10\Phi_0/L$ and the induced RF flux $\Phi(RF)$ is adjusted so that $\Phi(RF) \gtrsim Li_c$. In this case the critical current in the weak link is exceeded by the RF current and the flux quantum state of the ring can change by one or more flux quanta. Each time the quantum state of the ring changes, its energy changes so it is periodically perturbing the energy of the tank circuit. Thus as the quantum state of the superconducting ring changes it is supposed to traverse a hysteresis loop rather like those shown in Fig. II-1(a) and II-2(a) in synchronization with the RF excitation. The dissipated energy is approximately

$$\Delta E \cong 2\Phi_0 i_c - \Phi_0^2/L_2 \qquad \text{III-1}$$

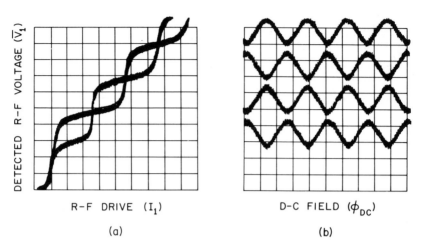

Fig. III-2(a). Detected RF voltage as a function of RF drive or bias current for two different values of DC applied flux differing by one-half flux quantum.

Fig. III-2(b). Detected RF voltage as a function of DC applied field for four optimized levels of RF bias current.

If $i_c \cong \Phi_o/L_2$, $\Delta E \cong \Phi_o^2/L_2$. Typically, $\Delta E \sim 10^{-30}/10^{-9} \sim 10^{-21}$ joules. The stored energy in the tank circuit is

$$E_1 \cong L_1 i_1^2 \cong \frac{1}{L_2} \frac{L_1 L_2}{M_{12}^2} \Phi_o^2 \qquad \text{III-2}$$

where the right hand equation holds if $i_c \cong \Phi_o/L_2$ and $E_1 \sim 10^{-19}$ joules. In this case each quantum state change makes about 1% change in the tank circuit energy lowering the Q to <100.

Zimmerman et al.[6] describe the consequence of a quantum transition of the superconducting ring as "shock excitation" of the tank circuit with gradual recovery of the tank circuit voltage over several cycles toward the original oscillation level until the quantum level of the superconducting ring again changes. The net effect is that the <u>average</u> RF oscillation voltage level is pulled to the levels at which the energy dissipated in the weak link and superconducting ring balances the RF energy input. This balance level depends on the magnitude of the DC flux applied to the ring. As RF drive current I_1 is increased the detected RF voltage increases monatonically but with a periodic modulation synchronized to the appearance of each additional voltage step that appears with each increase in the numbers of flux quanta driven in and out of the ring as shown in Fig. III-2(a) for two levels of applied DC flux.

The final fundamental question and indeed the crucial question, is how is the flux to be measured reflected in the RF voltage? That it is, is easily seen by comparing the two experimental curves shown in Fig. III-2(a) which correspond to DC flux of $n\Phi_o$ and $n+\frac{1}{2}\Phi_o$ where n is an integer. The differences between the levels of the plateaux provide a measure of the DC flux. In Fig. III-2(b) the effect of varying the DC applied flux at fixed optimum RF drive currents I_1 is shown. The triangular waves with flux period Φ have amplitudes corresponding to the spacing of the plateaux of the preceding figure.

In operating a flux detector the "DC" field is modulated at an audiofrequency and the RF current is constant. The RF voltage across the tank circuit is mixed with the oscillator signal in the first detector and the audio output is phase sensitively detected at the modulation frequency.

However, to display the properties of the flux detector it is useful to consider the RF voltage across the tank coil for various fixed DC flux levels with only the RF current amplitude modulated at an audiofrequency. In Fig. III-3, the audio modulated envelope of the RF voltage as observed with an oscilloscope operated with a linear time base sweep is displayed for several values of the DC flux. If the RF current were small enough the envelope would have been

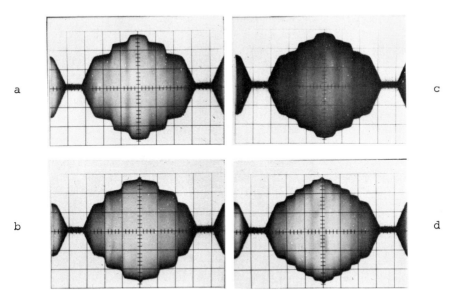

Fig. III-3. Oscilloscope traces of audiomodulated RF voltage across tank circuit with audiomodulated RF current at several levels of applied DC flux showing variation in plateau voltages with flux.

sinusoidal. However for this example, the RF current has been increased to the usual operating level. The current induced in the ring builds up until it exceeds i_c which has been chosen to be large enough to exclude several quanta, i.e., $i_c \gg \Phi_o/2L(1+\gamma)$ or assuming $\gamma \ll 1$, $i_c \gg \Phi_o/2L$. The pattern of plateaux at which the RF voltage tends to lock in as the RF current is swept by the audio modulation corresponds to the regions of small slope in Fig. III-2(a) and of course depends on the magnitude of the DC applied flux. If $\Phi(DC) = 0$ or $n\Phi_o$, the critical current is reached at a critical flux $\Phi_c \cong Li_c$ that is all to be supplied by the RF current, i_{1c}, for which the RF voltage is approximately

$$V_{1c} \cong \omega_o L_1 Li_c/M_{12} \qquad \left(\Phi(DC) = n\Phi_o\right) \qquad \text{III-3a}$$

where it has been assumed $\gamma \ll 1$ and $M_{12} \ll L_1$. Increasing $\Phi(DC)$ to a half-integer flux quantum of either polarity decreases the induced RF current needed for the total ring current to reach i_c and thus drops the first critical voltage to

$$V_{1c} = \omega_o L_1 Li_c/M_{12} - \omega_o L_1 \Phi_o/2M_{12} \qquad \left(\Phi(DC) = (n+\tfrac{1}{2})\Phi_o\right). \quad \text{III-3b}$$

The second term is the RF voltage reduction corresponding to a decrease of $\Phi_o/2$ in the induced RF flux in the superconducting ring. On reaching V_{1c} a flux quantum enters the ring extracting energy $\Delta E_1 \approx \Phi_o^2/L_2$ from the tank circuit, lowering the RF voltage by $\Delta V_1 \approx \cdot \tfrac{1}{2}(\Delta E_1/E_1)V_{1o}$, which by Eqs. III-2 and III-3 becomes

$$\Delta V_1 \approx \frac{(2\Phi_o i_c - \Phi_o^2/L)M_{12}}{\omega_o Li_c} \qquad . \qquad \text{III-4}$$

In general ΔV_1 is small and can be neglected in an elementary description since features associated with this voltage drop are generally hidden by fluctuation effects.

Successive plateaux appear at RF voltages corresponding to induced current changes of Φ_o/L if the applied DC flux just happens to be an integer or half integer flux quantum. Thus the maximum change of plateaux voltages with $\Phi(DC)$ is

$$\Delta V(RF) = \omega_o L_1 \Phi_o/2M_{12} \qquad . \qquad \text{III-5}$$

In the more general case where $\Phi(DC) \neq n\Phi_o$ and $\Phi(DC) \neq (n+\tfrac{1}{2})\Phi_o$ additional plateaux appear. For example, increasing $\Phi(DC)$ above $n\Phi_o$, each of the primary steps splits in half with the higher current half moving to higher RF voltage and the lower current part to lower voltage as shown in Fig. III-3(c). If $\Phi(DC) = (n+\tfrac{1}{4})\Phi_o$ the steps are again equally spaced at half the interval of the previous cases as in Fig. III-3(d). This result can be understood qualitatively with

MAGNETOMETERS AND INTERFERENCE DEVICES

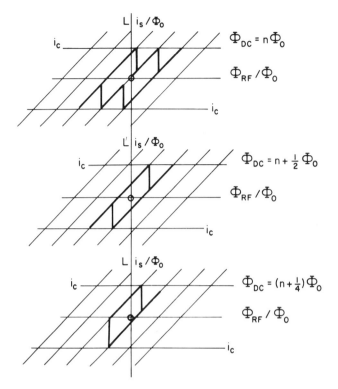

Fig. III-4. Hysteresis loops of flux detector ring for integer, half integer and quarter integer values of DC flux.

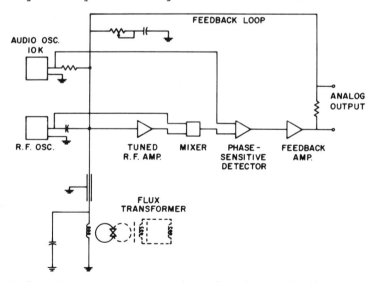

Fig. III-5. Flux sensor operating circuit-block diagram. Flux transformer (dotted) is discussed in a later section.

the help of Fig. III-4, which is an extension of Fig. II-1(c). Consider the total current in the ring as a function of the flux. The flux Φ_x now consists of two parts, the DC part to be measured and the RF bias part. The RF bias part cycles the current around hysteresis loops formed by the characteristics of neighboring quantum states and the transitions between them that occur whenever i_c is exceeded. The DC flux is taken into account in this diagram by a horizontal displacement of the quantum state curves. In the first case ($\Phi(DC) = 0$) a symmetrical hysteresis loop involving quantum states is traversed on every RF cycle by time the audio modulated RF level is highest at the re-entrant end of the plateau. Two additional quantum levels will be involved at the next plateau. In the second case ($\Phi(DC) = (n + \frac{1}{2})\Phi_o$) two quantum states are involved and the loop is symmetric. For the next plateau two additional quantum levels will be involved in the loop so the plateau spacing is the same as the previous case. However, in general, as in the third case, the hysteresis loop is asymmetric and only one more level is involved in expanding the loop to make the next plateau, so the plateaux are spaced half as far apart.

For actual measurements of flux the RF current is fixed at a value corresponding to one of the plateaux for $\Phi(DC) \neq n\Phi_o$ or $(n + \frac{1}{2})\Phi_o$ and an audiofrequency magnetic field of amplitude Φ_o is applied to the ring. The RF output detected at this audiofrequency then depends on the DC magnetic flux as follows: If $\Phi(DC) = 0$ or $n\Phi_o$ or $(n + \frac{1}{2})\Phi_o$ the output is zero at the modulation frequency while if $\Phi(DC) = (n + \frac{1}{4})\Phi_o$, it has a maximum. This result can be seen by the behavior with DC field of the plateaux previously discussed. At $\Phi(DC) = 0$, $n\Phi_o$ or $(n + \frac{1}{2})\Phi_o$, the plateau voltages change in the same direction with either sign of a change of flux since they are already at an extremum of position. Thus the output of frequency is double the modulation frequency and the phase depends on whether the DC flux is a whole or half integer. In any other case the direction of change of the plateau voltages do depend on the sign of the flux change. The detected output in this case reaches maximum amplitude at $\Phi(DC) = (n + \frac{1}{4})\Phi_o$.

To facilitate sensitive flux measurements, the circuit outlined by the block diagram in Fig. III-5 is used. The superconducting weak link, tank circuit and RF oscillator and audio field modulation have already been discussed. In addition the broad band RF amplifier amplifies the signal before it is mixed with an RF reference signal and the output is phase sensitively detected at the audio modulation frequency of the flux. Finally, in the usual technique the detector output is amplified and fed back through a direct coupled circuit to the inductance in the tank circuit. This signal provides a quasi-DC bias flux that maintains the RF output at an extremum. The feed back current is measured to provide a linear measure of the applied flux.

Another approach provides a lower sensitivity flux detector with a digital output count of quanta of applied flux. To do this the feed back circuit is eliminated. Instead the RF voltage is detected at the first harmonic of the audio modulation frequency as well as at the fundamental. Zero crossings of the two detected signals are fed to a counting circuit capable of reversible counting of the zero crossings with the sign of the count determined by the harmonic signal. This circuit developed by Forgacs and Warnick[7], has been used in many variations. Buhrman et al. at Cornell have used this kind of circuit continuously without "missing a count" for measurements of several days' duration in their measurements of nuclear magnetism as a milliKelvin thermometer.[8]

In summary, the superconducting quantum flux detector as presently realized consists of a weakly superconducting ring coupled to an L-C tank circuit resonant at 1 to 30 MHz with both at liquid helium temperature. The tank circuit is driven by an RF oscillator and an audiofrequency oscillator. The RF voltage across the tank circuit is amplified with a RF amplifier, mixed with a reference signal from the oscillator and phase sensitively detected with a lock-in detector at the flux audio modulation frequency. For high sensitivity, continuous, linear flux measurements, the output of the phase sensitive detector is amplified and fed back into the tank coil to provide a fixed operating point for the superconducting ring. The feed back signal is monitored to provide a linear measure of the applied flux. An alternative lower sensitivity digital flux quantum counting output is obtained by replacing the feed back circuit by an appropriate counting circuit.

IV. PRACTICAL DESIGN, OPERATION AND PROPERTIES OF FLUX DETECTORS

This section describes some of the practical problems that arise in design and operation of superconducting quantum flux detectors. Design and construction of the superconducting ring, shielding of interference noise and sensitivity limitations and alternative configurations are discussed.

A. Weakly Superconducting Rings

Fabrication of satisfactory weakly superconducting rings is still the most difficult problem in building superconducting flux detectors. These rings are required to have flux quantization and a critical current that remains stable and free of excess noise over long times and is insensitive to thermal cycling and variation of operating temperature. Adequate rings are, however, being fabricated

in several ways by at least six different groups at the present time. There are three schemes for fabrication that can be made to work reproducibly:
 (1) Machined massive ring structures that are closed by weak links formed by mechanical point contacts on the tip of adjustable screws; all fabricated of niobium.
 (2) Evaporated films of soft superconductors with reduced sections or bridges to form the weak link.
 (3) Electropolished ultrathin sputtered films of niobium with so far unknown weak link configurations.

These configurations are illustrated in Fig. IV-1 and discussed below.

Many investigators have worked on these problems and much of the art is unpublished so the literature cannot be cited here in detail. However the point contacts were first developed primarily by J. E. Zimmerman and Arnold Silver and their associates then at the Ford Motor Company with substantial success and later in the author's laboratory at Cornell University especially by M. R. Beasley and by R. A. Buhrman. Early use of isolated evaporated film bridges for research purposes is associated with A. H. Dayem at Bell Laboratories, and J. E. Mercereau and his associates at Ford Motor Company and now at Cal Tech have developed many configurations for flux detector use. They have been particularly interested in using the proximity effect to reduce the critical current in the bridge. The most recent development is the successful use of sputtered niobium films by John Goodkind at LaJolla. Although these contacts seem to be noisier than point contacts and the method still involves considerable art that has not yet generally diffused to other laboratories, this method appears most promising for the future.

The difficulties in fabricating flux detector rings arise from the requirements of the weak link which must have a critical current of about 10^{-6} to 10^{-5} amperes, requiring a cross-sectional area of $\leq 10^{-10}$ cm^2 which must be mechanically, chemically and thermally stable. Furthermore, this critical current must be reasonably insensitive to temperature. Some especially clean point contacts and the sputtered niobium films appear to have the necessary properties. These point contact structures are, however, bulky and expensive to fabricate, (see Fig. IV-1(a) (upper left)). and some groups have had difficulty with mechanical stability under thermal cycling. In order to maximize mechanical stability, the double ring illustrated in Fig. IV-1(b) is often used. (Both of these designs are due to J. E. Zimmerman and coworkers.) Here the applied flux is generated in a coil in one ring and the tank coil is inserted in the other. The flux then seems to slip back and forth across the point contact between the two rings. Various groups have found that, with protection from moisture and severe mechanical shock, they are stable for many years. The evaporated thin film structures are basically simpler but so far all suffer the severe limitation of strong operating tem-

MAGNETOMETERS AND INTERFERENCE DEVICES 669

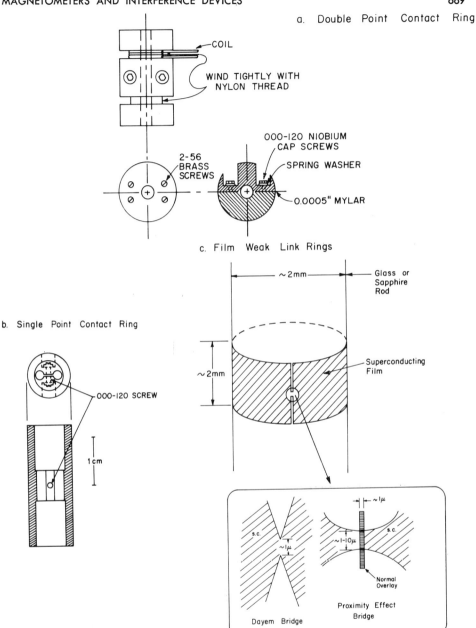

Fig. IV-1. At the top in (a) is the design of a double point contact ring which is convertable to a single link ring by simply setting one of the screws very tight; at the left in (b) is a very successful double contact ring design still frequently used. At the lower right two film type devices are shown, with details indicated in the inset.

perature dependence. Sputtered niobium films appear to be free of this limitation.

The details of the structure of the point contacts appear to introduce variations in the efficiency of their operation. If they are too wide, excess noise arises due to the trapping of flux quanta within the contact. If they are too long, thermal fluctuations are sometimes increased and the modulation amplitude of the critical current is decreased. The resistance of the weak link in the normal state is an important parameter and a large value, say $R_n > 1$ ohm, is preferred. The junction capacitance must be sufficiently small that $(RC)^2 2ei_J/\hbar C < 1$ to avoid hysteresis. However, it is not yet clear what is the optimum capacitance. Conventional large area Josephson junctions made by evaporation of a superconducting film on top of the insulating oxide formed on another superconducting film to form a superconductor-insulator-superconductor sandwich are not suitable for flux detectors because of this excess capacitance. Fig. IV-1(c) shows the essential features of the evaporated film flux detector configurations. At the top, the usual overall geometry is seen to consist of a cylindrical evaporated film deposited on an insulated substrate. Usually the film has a reduced section containing a weak link. Two forms of weak link, the Dayem bridge and the Mercereau proximity effect bridge are shown in the inset. The Dayem bridges can be made as small as 1 μm wide × 1 μm long × 0.5 μm thick using optical photo-resist or conventional masking techniques. Using electron beam sensitized photo resist techniques, they have been made substantially narrower. Except for the previously mentioned niobium film weak links, all of the available thin film weak links appear to suffer from a narrow range of operating temperatures.

B. Shielding of Interference

Flux detectors are, of course, extremely sensitive to the magnetic fields in which they are operated. Furthermore, they are sensitive to radio frequency interference, as readers of Professor Shapiro's chapter on radiation detection will realize. Thus flux detectors must be shielded from both.

Electrostatic shielding of all electrical circuits entering the dewar that might act as antennas is necessary, but not sufficient. Careful filtering and isolation of all extraneous circuits is essential to low noise performance. The electronic noise of the amplifier and sometimes the oscillator circuit probably limits the effective sensitivity of flux detectors. The signal to be measured with the flux detector may also carry unwanted magnetic noise. To filter this noise, flux detectors in the author's laboratory are sometimes fitted with a thin copper or brass shield designed to screen out field changes above a cut off frequency that can be set by choice of the wall thickness and conductivity for the appropriate eddy current shielding. This kind of shielding has also been used to isolate the RF bias signal from sensitive specimens in the magnetometer.

Shielding of ambient magnetic fields is necessary unless the flux detector is to measure the field, and in that case it is still useful to shield the flux detector from all fields except the flux introduced by a suitable pick up coil or "flux transformer" that will be discussed. This shielding is facilitated by use of the two hole flux detector ring illustrated in Fig. IV-1(b), since it is a gradiometer and is insensitive in principle to uniform fields. However it is usual to use two or more layers of "permalloy" or "mu-metal" shielding outside of the dewar and a superconducting can inside where feasible. A superconducting lead can at helium temperature completes the shielding, and fields of $<10^{-5}$ Gauss that are steady to 10^{-9} Gauss can be obtained with reasonable ease. Intrinsic fluctuations in the earth's field are much smaller than the field fluctuations due to man's vehicles and power sources, particularly diesel electric locomotives and elevators. Unless it is magnetic anomalies that are to be observed, the flux detector must be either shielded or isolated from unwanted disturbances.

C. Alternative Configurations of Flux Detectors

Two schemes for superconducting quantum flux detectors which differ substantially from those described here have been in wide use. They both consist, in essence, of superconducting rings with two weak links. In this case the bias current is usually a DC or audio frequency current; sometimes the two are superimposed. The analysis is closely analogous to an optical two slit interference pattern so one form of this device is called a superconducting quantum interference device for which the acronym is SQUID. This term is often applied to the RF biased single point contact device as well -- although it is not strictly an interference device. Zimmerman and Silver described an effective version of the two-contact SQUID in 1966. A version developed in the author's laboratory has proven convenient and has been widely used.[10] A different form of the multiple link device was developed by John Clarke and has been and is used effectively by him, particularly for voltage measurements.[11] This device is constructed by melting a blob of lead-tin solder around an oxidized niobium wire. Due to its appearance, it carries the acronym SLUG.

D. Sensitivity and Signal-to-Noise

The noise limit of the flux detector is not yet fully established. The flux detector signal power, i.e. the sideband power in the modulated RF response, is $\sim (\Phi_o^2/2L)(\omega_o/2\pi)$ so the signal voltage is proportional to $L^{-\frac{1}{2}}$. This result follows from the value of energy of a fluxoid in the loop, $\frac{1}{2}\Phi_o^2/L$, which is approximately the maximum energy extracted from the tank circuit each cycle. If the electronic noise of the flux meter electronics dominates the instrument, it can be expressed as an equivalent noise current δI_M in the magnetometer inductance with the help of the Nyquist theorem: $(\delta I_M)^2 = 4kT\delta\nu/R_M$ where R_M is an equivalent noise resistance. (Alternatively, we could define a noise temperature, T_M.) The equivalent noise voltage in this case can be written $\delta V_M = 4kTR_M\delta\nu$, where $\delta\nu$ is the instrument time constant. The signal-to-noise voltage ratio would then be proportional to $L^{-\frac{1}{2}}$, suggesting a minimum value of L for maximum S/N.

If, on the other hand, the intrinsic thermal noise in the superconducting ring dominates, the noise is $<\delta\Phi^2> = LkT$ and the Nyquist theorem gives the classical flux noise spectral density as $<\delta\Phi^2>/\delta\nu = 4LkT(L/R)/(1 + \omega^2 L^2/R^2)$, which becomes, at low frequencies,

$$<\delta\Phi^2>_\omega = 4L^2 kT\delta\omega/R \qquad . \qquad \text{IV-1}$$

Typically $L \sim 10^{-9}$ H, $R \sim 1\Omega$ so $(<\delta\Phi^2>/\delta\omega)^{\frac{1}{2}} \sim 10^{-20}$ Weber/Hz$^{\frac{1}{2}}$ or $\sim 10^{-4}$ to 10^{-5} flux quanta. However it is not at all clear that this classical calculation is appropriate for this problem, since the superconducting ring is actually being switched amongst quantum states by the modulated bias flux and the RF bias. One calculation by J. E. Lukens which takes this mechanism into account yields a still lower fluctuation level.

The observed flux fluctuations in practical operating flux sensors are typically $\sim 10^{-3}$ flux quanta, which suggests that the actual noise is limited by the electronics, probably the RF amplifier. Thus there appears to be significant room for improvement. There are, however, not yet any experiments that definitely establish the origin of the observed fluctuations or that confirm any particular theoretical limit. Again, for practical purposes, the flux sensitivity is usually about 10^{-3} flux quanta, with a one second averaging time, i.e. 2×10^{-10} Gauss cm^2 sec$^{-\frac{1}{2}}$.

V. MEASURING INSTRUMENTS

The superconducting quantum flux detector provides the basis for measurements of magnetic fields or magnetic susceptibility and for electrical measurements of voltage, currents and resistances. Configurations of these instruments are outlined here.

For all instruments, the key to flexible design is a handy accessory based on the perfect diamagnetism of a closed superconducting ring that is called a flux transformer. A superconducting wire is wound into two coils and the ends joined into a closed loop to make a flux transformer. If an applied field is applied to one coil, a supercurrent is induced that compensates for the applied field. The compensation is not perfect because the induced supercurrent also produces a field in the second coil. The criterion is simply that the total flux threading the closed superconducting loop remain constant. This simple device provides a way of transferring magnetic flux signals with an efficiency approaching, but not exceeding, one-half.

To use a flux transformer, one coil is coupled into the weak superconducting ring as tightly as possible and the other coil of the closed loop is positioned to intercept the flux to be measured. (See dotted portion of schematic diagram, Fig. III-4.) Any superconducting wire -- tin, lead, niobium or even tin coated copper -- works quite well as a superconducting flux transformer in small magnetic fields. For magnetization measurements, some of the coils are wrapped around a specimen and some are inserted in the flux detector. A magnetic field is applied using a superconducting magnet in the persistent current mode and the magnetization is measured by observing the change of flux on removing the specimen.

In magnetic field or susceptibility measurements, appropriate design of the flux transformer increases field sensitivity without increasing magnetometer noise. However, the coupling coefficient between the flux transformer and the weakly superconducting ring is difficult to raise above about 0.8, which limits its efficiency. Zimmerman[12] has developed an approach in which the weakly superconducting ring is divided into several uncoupled loops connected directly across the weak link in parallel to increase field sensitivity at minimum noise level, and this scheme appears to yield somewhat better field sensitivity than a flux transformer. At the present time magnetic field sensitivity of $\sim 10^{-9}$ Gauss at a signal to noise ratio of one with a one second time constant measurement has been reached in many laboratories and may be considered routinely obtainable. It is generally agreed that substantially lower noise will be achieved in the future but there is no concensus about the means nor the magnitude of the maximum possible improvement.

Measurements of voltage, resistance and current with a superconducting quantum flux detector are readily accomplished by coupling a "multimeter" circuit into the flux detector with a flux transformer to make a very sensitive low frequency amplifier. Simplified diagrams are shown in Fig. V-1 for a voltmeter and an ohmmeter as developed in the author's laboratory primarily by J. E. Lukens.[13] The voltmeter circuit has an extremely low impedance, of course, but this is not a prohibitive disadvantage, since many sources of very small voltages are low impedance sources. For example, voltages from $\sim 10^{-6}$ ohm sources can be measured with a one second time constant. The measurement power of these instruments is $\leq 10^{-22}$ watts with a one

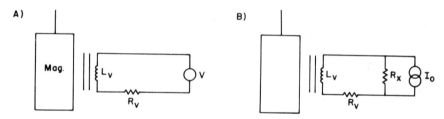

Fig. V-1. Multimeter circuit based on quantum flux sensor. In (A) small voltages are measured by observing the flux generated in L_V by the current driven through R_V and L_V by the unknown V. In (B) the analogous ohmmeter circuit is shown.

second time constant. Notice that the thermal energy $k_B T$ at liquid helium temperature is $\sim 10^{-22}$ joules. In some applications, it is desireable to obtain a large input impedance and this can readily be done at some loss of sensitivity by operation in a potentiometric mode with feedback into the ohmmeter or voltmeter circuit to balance out the input current.

An apparatus used for this kind of measurement is basically simple but becomes complicated by the precautions required to exclude extraneous perturbations at these extremely low signal levels. A schematic diagram of the arrangement used by Lukens, Warburton and Webb is shown in Fig. V-2 to illustrate the ingredients. The small can at the upper left labeled magnetometer is the crucial device described in this chapter, but the remainder of the apparatus which provides shielding and thermal stability was just as necessary to insure valid measurements.

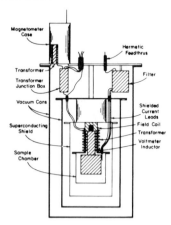

Fig. V-2. Schematic diagram of the structure of the superconducting quantum multimeter. Most of the complexity is required to accommodate accessories. The flux detector occupies the relatively small lead plated can at top.

VI. APPLICATIONS AND FORECASTS

The capability of superconducting quantum flux detectors has been exploited for several years in measurements of the properties of superconductors. Flux creep[14], the onset of resistance in one dimensional superconductors[15], fluctuations in weakly superconducting rings[16] and diamagnetism above the critical temperature[17] are well known examples.

Superconducting flux detectors are utilized in two schemes for thermometers at low temperatures. Kamper and Zimmerman[18] have worked on schemes for measuring thermal voltage fluctuations in a resistor as a measure of temperature in the liquid helium region, and Buhrman et al.[8] in the author's laboratory have measured the magnetic susceptibility of metallic nuclei, which is supposed to vary as T^{-1} to temperatures as low as $T \sim 10^{-5}$K, as a measure of temperature in the milliKelvin region. Other applications of this sort have been made by Wheatley and his coworkers and by Goodkind and his coworkers at La Jolla.

Outside of solid state physics, promising applications appear in geophysics, particularly in rock magnetism where the superconducting magnetometer makes possible the study of weakly magnetic sedimentary rocks as well as the igneous rocks usually studied in determining the magnetic history of the earth. In medical applications, striking successes have been obtained by David Cohen and coworkers who are studying the behavior of the mammalian heart using a superconducting magnetometer in magnetocardiography, a technique which has already demonstrated its effectiveness as a research and diagnostic tool.[19]

Instruments are now commercially available from at least three sources at costs beginning at a few thousand dollars.[20] Portable flux detectors complete with reliable cryostats which hold a 12 hours' supply of helium, yet small enough to fit into a brief case, have been built and operated. The weakly superconducting ring in these instruments apparently requires occasional adjustment or replacement, but this minor disadvantage should soon be overcome commercially as it has been in laboratories.

It appears completely feasible to build certain instruments for field use and for untended operation over times of the order of weeks -- limited only by helium supply. However, quantum superconducting flux detectors are inherently very sensitive. They cannot now be expected to attract widespread routine use because the sensitivity is seldom needed. The importance and difficulty of screening interference limits them to circumstances where their special sensitivity applications may be facilitated by the convenient digital nature of fluxoid counting mode or operation. Flux detectors may be applicable in superconducting computer circuitry in which hysteretic

Josephson junctions provide the memory elements as described in the chapter by J. Matisoo. In any case, it seems likely that some low level applications that are not yet recognized will arise, although their nature is not presently obvious.

VII. ACKNOWLEDGMENTS

It is a pleasure to acknowledge the sponsorship of research on superconducting sensors in the author's laboratory by the Office of Naval Research and by the Advanced Research Projects Agency through the Materials Science Center at Cornell, and to thank Dr. J. E. Lukens for numerous helpful suggestions and corrections. Much of the analysis of weakly superconducting rings follows the original work of A. H. Silver and J. E. Zimmerman, and permission to reproduce several of their figures is acknowledged with appreciation.

VIII. REFERENCES

1. A. H. Silver and J. E. Zimmerman, Phys. Rev. <u>157</u>, 317 (1967).

2. R. DeBruyn Ouboter and A. Th. A. M. DeWaele in Progress in Low Temperature Physics VI, C. J. Gorter ed. (North Holland Publishing Co., Amsterdam, 1970), p. 243.

3. P. W. Anderson in Progress in Low Temperature Physics V, C. J. Gorter ed. (North Holland Publishing Co., Amsterdam, 1970).

4. Using the Ginzburg-Landau equations for the current-phase relation at small current in a thin wire gives $\theta = (2\pi/\Phi_0)(m/e^2 n_s)(\ell/\sigma) i_s = (2/3\sqrt{3})\left[\ell/\xi(T)\right] i_s/i_c$ where $\xi(T)$ is the superconducting coherence length and i_c is the critical current. It is interesting to note that setting the link length $\ell = (3\sqrt{3}/2)\xi(T)$ and i_c and i_J gives the low current phase relation corresponding to the Josephson equation.

5. J. M. Goodkind and D. L. Stolfa, Rev. Sci. Instr. <u>41</u>, 799 (1970).

6. J. E. Zimmerman, P. Thiene and J. T. Harding, J. Appl. Phys. <u>41</u>, 1572 (1970).

7. R. L. Forgacs and A. Warnick, Rev. Sci. Instr. <u>38</u>, 214 (1967).

8. R. A. Buhrman, W. P. Halperin, S. W. Schwenterly, J. Reppy, R. C. Richardson and W. W. Webb, Proc. 12th Intern. Conf. on Low Temp. Phys., Kyoto 1970.

9. J. E. Zimmerman and A. H. Silver, Phys. Rev. <u>141</u>, 367 (1966).

10. M. R. Beasley and W. W. Webb, Symposium of the Phys. of Superconducting Devices, University of Virginia, Charlottesville 1967.

11. J. Clarke, Phil. Mag. 13, 115 (1966).

12. J. E. Zimmerman, to be published.

13. J. E. Lukens, R. J. Warburton and W. W. Webb, J. Appl. Phys. 42, 27 (1971).

14. M. R. Beasley, R. Labusch and W. W. Webb, Phys. Rev. 181, 682 (1969).

15. J. E. Lukens, R. J. Warburton and W. W. Webb, Phys. Rev. Letters 25, 1180 (1970).

16. J. E. Lukens and J. Goodkind, Phys. Rev. Letters 20, 1363 (1968).

17. J. P. Gollub, M. R. Beasley, R. S. Newbower and M. Tinkham, Phys. Rev. Letters 22, 1288 (1969).

18. R. A. Kamper and J. E. Zimmerman, J. Appl. Phys. 42, 132 (1971).

19. D. Cohen, E. A. Edelsack and J. E. Zimmerman, Appl. Phys. Letters 16, 278 (1970); D. Cohen, J. C. Norman, F. Molokhia, W. Hood, Jr., Science 172, 1329 (1971).

20. Suppliers known to the author are S.H.E. Corporation, La Jolla, Calif.; Develco Inc., Mountiain View, Calif.; Macroquan Data Systems, Inc., Los Angeles, Calif.

Part IV
Panel Discussion
The Scientific, Technological, and Economic Implications of Advances in Superconductivity

PANEL DISCUSSION: THE SCIENTIFIC, TECHNOLOGICAL, AND ECONOMIC IMPLICATIONS OF ADVANCES IN SUPERCONDUCTIVITY

W.N. Mathews Jr.

Edited by   W.D. Gregory

E.A. Edelsack

    The invited speakers of the Georgetown University short course on "The Science and Technology of Superconductivity" encompass extensive and diverse backgrounds in the field of superconductivity. Their combined presence at Georgetown University afforded a singular historic opportunity to solicit and record their views on the scientific, technological, and economic implications of recent advances in superconductivity. The first hour of the discussion centered on the implications of the major scientific advances which occurred in the last five years in superconductivity research. The second hour of the discussion focused on the technological and economic implications of the applications of superconductivity during that same five year period.

    The discussions were recorded on video tape. Their viewing provides a most stimulating panoramic view of the opinions and ideas of twenty-two leading workers in the field of superconductivity. These video tapes, as well as those of the individual speakers, are available to those individuals interested in viewing the highlights of the course. For those unable to view the tapes, it was decided to transcribe the entire discussion. The verbatim transcriptions have been lightly edited to improve their readability. The discussants were given an opportunity to revise their remarks, and several did. However, any errors or shortcomings in these edited transcriptions of the discussion fall solely on the shoulders of the editors.

The following invited speakers took part in the discussion:

J. Bostock	A. Martinelli
H.T. Coffey	W.N. Mathews Jr.
B.S. Deaver, Jr.	J. Matisoo
T.F. Finnegan	B.T. Matthias
W.D. Gregory	R.M. Rose
D.A. Haid	D.J. Scalapino
W.O. Hamilton	J.F. Schooley
R.A. Hein	S. Shapiro
C.S. Koonce	J.L. Smith, Jr.
D. Langenberg	Z.J.J. Stekly
M.S. McAshan	W.W. Webb

The moderator was E.A. Edelsack.

Edelsack:

Let us begin by considering theoretical research in superconductivity. Theoretical research can be divided into three parts: that which directly relates to the properties of superconducting materials; that which is concerned with the properties of superconducting devices; and that which does not directly relate to devices or materials. An example of the latter might be research on the BCS theory. Dr. Scalapino, within this arbitrary categorization, what have been some of the important advances in theoretical research in superconductivity within the last five years. And, if I might add, are there any discernible trends for the future?

Scalapino:

As you know, things get telescoped, so that sometimes what took place in 1962 looks to me now like it took place two years ago. I'll try to avoid that. Let me see if I can take what you asked in order. The first area was materials. The strong-coupling theory of superconductivity, in which the details of the electron-phonon interaction are taken into account, has allowed us to correlate certain data to quite a high accuracy. It leads us to the conclusion that, for the materials which to date have been analyzed, the electron-phonon interaction is responsible for superconductivity. For example, the recent tunneling data for tantalum can be used to calculate T_c for this material to within several percent. In fact, all of the strong-coupling theory, when it's applied, has usually been within a few percent. This, of course, arises because of a subtraction procedure in which we calculate superconducting properties relative to the normal state properties, but it leads us to have great faith in our picture of the basic process. Furthermore, it means that, given tunneling data, we can successfully calculate other properties. How you calculate with limited and more limited data is still a question. In line with the strong-coupling theory, the other subject I would like to raise now is McMillan's work on the transition temperatures.

McMillan has attempted to calculate and predict transition temperatures using strong-coupling theory. This work has shown what limits are possible within phonon systems. It has led to an understanding of how softening of the phonon spectrum in thin films can increase the transition temperature. It has given us a remarkably accurate description of events in terms of some limited parameters. Some of these parameters are more difficult to obtain than others. Some are not as simple as the Debye temperature, but are certain averages over the phonon spectrum weighted by an effective frequency-dependent interaction, which means one needs rather great detail. However, this work further supports the idea that we do have the fundamental theory in hand as far as metals go. Further, it seems to me that the area of transition metals is also falling into place. I would not like to claim that it is fully understood, but I would like to claim that we now, with the work of Schrieffer and co-workers, understand the relationships between things like specific heat, susceptibility, transition temperature, resistivity, etc., that we did not understand before. Finally, with the recent work of Sham we are beginning to get an understanding of what the soft modes are in the β-W structures.

Your second area had to do with devices. Here it's difficult for me because I personally think that some of the important advances in the device area are the currently evolving equivalent circuit representations for various weak link phenomena, but I would not classify these as theoretical; I would classify this work more as engineering and as the experimental adaption of results to fit certain parameters. I would say that there has been an increased understanding on the theorist's part of noise in the weak links and of the ultimate limitations of weak link devices in circuits.

Your last area includes everything that doesn't fall into the first two areas. Off hand, I would pick the area of fluctuations. One of the things that prompted the work on fluctuations was consideration of one- and two-dimensional superconductors in which the following questions were raised: What do we mean by superconductivity? What do we mean by the phase transitions? Can we have systems that are almost superconducting? That is, can we have systems that for practical purposes have zero resistance and yet have not undergone a phase transition? Can we have systems, therefore, that do not undergo phase transitions because of either their dimensionality or their restricted size and yet have some of the properties of bulk superconductors? Out of these ideas have come some very definite theoretical predictions. The most striking of these, I'm sure, would be the Aslamazov-Larkin prediction of the appearance of an anomalous contribution to the conductivity above T_c. This said to us: "Look, this phase transition is not so different from the magnetic transition in which the Curie-Weiss behavior of the magnetic susceptibility warns us of a possible phase transition." Superconductivity, you may remember, at least if you go back more than five years, was always a bit spooky in the sense that you didn't see it coming until it hit you. What's been understood now in terms of the fluctuation theories is that there are well defined precursors of the superconducting phase transition.

Very recently work on two-dimensional tantalum sulfides, intercalated type compounds, gives evidence for superconductivity in the behavior of the diamagnetic susceptibility up to 40 or 50 K.

Along with fluctuations, I would like to mention one other area, that of metastability. This has been with us a long time in superconductivity, since Bloch first proved that the ground state of a loop is not a current-carrying state. How about currents in loops which persist for as long as the loops are kept cold? The work by Langer and Ambegaokar has shown how fluctuations could give rise to the decay of the metastable states.

You ask for descernible trends, but I'm not sure but what I've talked too long and maybe should stop.

Edelsack:

No, please don't. Suppose a graduate student came to you and said, "I want a good problem in theory to tackle." What would you suggest?

Scalapino:

I personally feel that there's still interest in this question of systems that are nearly superconducting. There are, in fact, exact solutions now becoming available for one-dimensional and zero-dimensional superconductors. By zero-dimensional I mean a particle whose diameter is smaller than the Ginzburg-Landau coherence length. This restricts the fluctuations that dominate the phase transitions to uniform fluctuations of the order parameter. With this restriction one can actually carry through complete calculations of the thermodynamic properties of these systems. As these systems become large you find, for example

PANEL DISCUSSION 685

a sharpening of the calculated specific heat jump, and as they become smaller and smaller you see how this blurs.

Edelsack:

Dr. Bostock, do you have any comments? What are your own views as a theorist?

Bostock:

Well, as far as trends for the future go, I think Dr. Rose probably convinced many of us that there's going to have to be a lot of work done on Type II superconductivity and vortex flow. They certainly are doing this work in Europe. The theoretical explanation of what's going on in amorphous films is also a very important problem being considered there.

Edelsack:

Dr. Mathews, what would you have to say?

Mathews:

I can only add a few things to what Dr. Scalapino and Dr. Bostock have said. In the area of materials work that has been going on and, of course, is going on now, I think the work on A-15 materials is significant. These are, by and large, the highest critical temperature materials and an understanding of them will undoubtedly lead to a better understanding of superconductivity as such, and perhaps to other things. But we'll let Dr. Matthias talk about them. I think that some of the theoretical work that has led to a better understanding of Type II materials in general is also significant. The work of Bardeen, Kümmel, Jacobs, and Tewordt, for example, would be included in that. Then finally, I mention some of the work on looking at other mechanisms. For example, Rothwarf's work on plasmons, which recently appeared. Some of the spin wave work and perhaps, I hesitate here, perhaps some of the work in which excitons are considered is of significance too. I would include this with the work on materials because specific mechanisms other than the electron-phonon interaction may be important for particular materials, rather than in general. I think that I have to agree with Dr. Scalapino that, so far, the electron-phonon model has accounted for, in what must be regarded as impressive detail, much of the experimental behavior of superconductors. But the other mechanisms have to be either laid to rest, or at least considered. It's not at all clear that if there are other mechanisms beyond the electron-phonon interaction the isotope effect would come about via a unique combination. Perhaps more than one combination of these same interactions would lead to a given isotope effect. This is certainly, of course, also a subject for future research.

In the area of devices, I think many of the most significant devices being considered today, perhaps a majority, depend upon the Josephson effect, or weak link behavior, in one way or another Thus the theoretical work that's been done on the Josephson effect in the last few years is highly significant. I would particularly cite some of the work that Werthamer has done in connection with Josephson tunnel junctions.

Now as far as other theoretical work, I can only mention three things. The work on space-dependent effects in superconductivity - the intermediate state, the mixed state and that sort of thing. And then the work on time-dependent superconductivity, and I have to say that Professor Scalapino's talk yesterday was one of the most exciting developments I've heard recently. And then, finally, I think Eilenberger's transport equations. He essentially integrates over the Green's function equations and integrates out the variables that are not of immediate macroscopic significance to obtain somewhat simpler equations. This has to be regarded as a significant theoretical development, particularly since it may lead to a better understanding of transport processes in Type II materials. Now as far as trends, it's difficult to really be specific, but I can cite two things that are really connected. First, the work on mechanisms that I already spoke of. I think that this may contain some possibilities. If nothing else, some of these considerations will lay that problem to rest eventually, I hope. And then, finally, in much the same vein, I think that there remains work to be done in correlating the electron and phonon properties with superconducting properties.

Edelsack:

Dr. Koonce, since obviously the theoreticians are having their say first, would you have any comments that you'd like to make?

Koonce:

I think the field has been very well covered by Dr. Scalapino and the others. I would just like to emphasize once again the tunneling work on tantalum which has, I believe, shown that the phonon mechanism is primarily responsible for superconductivity in tantalum, and by implication in other transition metals. And I think that the work on the calculation of superconducting properties is a very important field. There has been a certain amount of work in the last five years in this field and I think it's going to be an important field in the future. I think this is, of course, closely related to what Dr. Mathews was saying about mechanisms for superconductivity.

Edelsack:

Now I'd like to turn to the experimental side of the house. Have these recent advances in theoretical research in superconductivity, which we've just heard about, been of any help to experimentalists working with materials or devices?

Webb:

I'd like to take the chance to make one more remark about theory. There's one area that the theorists have not mentioned, perhaps because it's an area where the theory is generally phenomenological. That's our understanding of that feature of superconductivity which makes it sell these days, the high critical current, which depends upon metallurgical inhomogenieties. The phenomenological theory of the so-called critical state, the pinning, the concepts started by Phil Anderson about ten years ago, have been extended by Kim, Bardeen, Stephen, Vinen, and others. That work is very much used in understanding hard superconductors. I think there's obviously more to do in this area. I notice that most of the work has not been done in the States.

Edelsack:

That's a significant point. I'd like to try to come back to that later, vis-a-vis what's going on in the States and abroad. Can we turn to the experimentalists in the discussion panel or in the audience? What help, if any, have these recent theoretical advances been to experimental work on materials and devices?

Deaver:

I'd like to repeat again that what Dr. Scalapino has done is very interesting and enlightening to those of us who for a long time had to think pictorially about how junctions and weak links are more or less alike and why we're able to make equivalent circuit representations of them. It's nice to see somebody take seriously the problem of solving what happens in time at least. I'd like to see it done in time and space to widen the thing out and to see the dynamics of the superconducting state. And this, to me, is a very exciting development.

Edelsack:

Dr. Shapiro, would you care to make some comments on where

the recent advances in theoretical research have been of utility to you?

Shapiro:

Well, Dr. Scalapino has thrown it out when he says it's engineering, although significant. Perhaps it is engineering, but it certainly is significant. The reality of the equivalent circuit models and the conviction that what you see there has some bearing on what you see in the real world has been occupying the minds and interests of many of us working in the Josephson device area. It has influenced our efforts on the ac properties. I think, as I believe Dr. Deaver has been implying, that it has also influenced our efforts on the low frequency or dc devices.

Edelsack:

In device research on the upper frequency limit and ultimate sensitivity of operation, do you feel that theoretical work in this area has been of any import in terms of your own experimental efforts?

Shapiro:

Not yet.

Edelsack:

Why? Are you limited by the instrumentation that you attach to your superconducting device or are you limited by the intrinsic noise properties of the device?

Shapiro:

Those are two different questions. Let me say why. The question that I'm interested in primarily is what you can do in a given system with these devices, and what you can do in a given system now is constrained by lack of serious attention to things other than the device. The numbers that people talk about in terms of ultimate sensitivity are extrapolations in which they try to take into account all the inefficiencies of the given system in order to reach a number that may have some significance for the device itself in isolation. I don't particularly feel that those numbers have very much significance. I don't think that they are particularly good guidelines for theorists and I'm sure that they're of no concern to someone who has a real problem involving detecting microwave radiation. That's why these considerations haven't yet had a great bearing on my research. From a theoretical point of view, the ultimate sensitivity may be an intriguing question, but in terms of developing better detectors,

I don't think that it is particularly helpful at this time.

Gregory:

There's an area which has not yet been mentioned and where we've recently found some help from the theoretical people. I refer to the selection rules for tunneling. There's been a lot of work done recently on tunneling into bulk single crystal materials. A year or so ago Dowman, MacVicar, and Waldram proposed a theoretical selection rule, which we recently found to work. The significance of it as a theoretical contribution is that there exists lots of data that could be very useful. You can look at the energy gaps, the variations as a function of crystal orientation and, if you have the correct selection rule, work your way backwards to find out something about the interaction. I think there should be more work in this particular area.

Schooley:

One area which has been mentioned is McMillan's calculations on transition temperatures for various classes of materials. My own viewpoint is that the results should be viewed as an upper limit in transition temperature to which one can aspire, if natural roadblocks don't intervene, for example, in the attempt to change the phonon frequency, and the resulting material, such as a film, doesn't suffer from other problems that weren't accounted for by the theory. McMillan's work has indicated the maximum transition temperature one could expect to see with the electron-phonon interaction for the transition metal superconductors. Experimental effort can then, with some justification, be redirected towards other areas or toward other possible interactions to try to achieve higher transition temperatures. So theory has helped in that sense. It may not be an entirely positive sense, but it is, I believe, a factor in making experimental progress.

Hein:

I think Schooley's remarks answer one of your initial questions; that is, we're talking about a theory which came out, at least among preprint circuits, in 1965 or 1966. It indicates to me that there hasn't been much theoretical progress made since 1966. Now McMillan's theory is used because we experimentalists can plug in numbers. I don't know of any theory that has come out since McMillan's theory that is used by the experimentalists. I think McMillan has been hammered a little too hard here. What he does is nothing more than what experimentalists do. He has an equation, he takes numbers, he puts them in, he calculates λ, and he extrapolates. The 30 K which he obtains is simply an extrapolation, which is just as valid as an extrapolation to a tenth of a milligree, going the other way.

Scalapino:

I wish to comment on this briefly. I think there have been new developments since McMillan's work. Prior to it, there was Garland's work on the various isotope shifts, but let me go to more recent things. After McMillan's work, people tried to further parameterize the phonon spectrum using other things besides average phonon energies and coupling strengths. I think most of this work simply introduced more parameters and was not any more successful than the original work. But I think something very significant then took place. From a very personal point of view, I'll describe this. Charles Owen was a graduate student of mine at the University of Pennsylvania. As part of his thesis research, we were looking for excitons in superconductors, pairs of quasiparticles binding below the gap. We didn't find them. But as part of this we developed a matrix technique, based on the strong-coupling theory, for calculating T_c. We took data from Dynes and Rowell for elements, two- and three-element compounds, and amorphous systems, ran it through, and always found T_c's within 5%, sometimes within 2% of what was experimentally observed. Out of this comes the idea that we really have to trust the strong-coupling theory to a high degree of accuracy. So we went back and looked at McMillan's formula. Dynes had then published an article in which he has used McMillan's formula with 1.45 θ_D replaced by an ω average. We did the same thing and got very different answers. We wrote Dynes a letter about this. It turns out that Dynes had made an algebraic error. He later published an errata. I think this is a most significant new piece of work. It says the following: if you take McMillan's formula and use the following very strange average (but for good reason) - you take the phonon spectrum and integrate it to give you a numerator, and down below you take the phonon spectrum and divide by $\frac{1}{\omega}$ and integrate it - in place of McMillan's θ, you get excellent agreement for all of the systems on which Dynes has tunneling data. At any rate, the McMillan theory is excellent if you use this reduced ω. The meaning is that if you drop the phonon frequency, so as to enhance the coupling constant which goes in the exponential, you've got to pay for it. And the place you pay for it is out front, you have to use a lower θ_D out front. I think you will find that the idea fits very nicely for bismuth-lead alloys - how the T_c's should vary. I think that the theory here has done very well and can in fact be used within these systems to predict.

Edelsack:

Phil Anderson said some time ago that, in terms of T_c, it seems definite that ordinary metals and orthodox phonon mechanisms cannot carry us much further. The phonon mechanism gives us a definite upper limit for the transition temperature, not very much above the values already achieved. Dr. Matthias, would you care

PANEL DISCUSSION

to comment on this viewpoint?

Matthias:

There was a reporter by the name of Plum, who, when he announced the discovery of BCS in the New York Times, said "We have been assured by the authors that in the near future the superconducting transition temperature will be raised considerably." That was in 1957, and listening to my friend Scalapino, nothing seems to have changed during the last fourteen years. We're still assured everything is good and well. We understand it. We have more parameters. We interpolate more, extrapolate less. What does it get us? What has been achieved? Well it's this. A group of aging physicists like us have raised the transition temperature finally another 3 Kelvins from 18 to 21 K, and a group in Los Alamos has discovered a new system that leads to 17 or 18 K. Apart from this, what new materials, what new transition temperatures have been accomplished? Not one single discovery. To come to McMillan's theory, I was there when he wrote it and I saw the resultant redirection of research. I pleaded with the people who were being redirected not to do it, that it was going to be a waste of time and money. And three years later, it turned out to be. Nothing came of it. Sure, McMillan has given new higher limits of 28 K, 40 K, but has anybody raised anything on this basis? No. It has led to many new experiments, all of them negative. How anybody can call this an achievement is beyond me. How come from all this no new materials have come? Surely, it's true. You can fit everything, you can calculate everything. But can you predict? You must describe old data. On that day when you suggest an experiment and it leads to a result, only then I'd say we really do understand things profoundly. Now you mentioned a few thing about mechanisms and how they should be laid to rest. Let me just remind you what the new mechanisms have accomplished. When we realized that there was a new mechanism in lanthanum, we used this new mechanism and we raised the transition temperature from 6 K to 12.7 K. In other words, we more than doubled the transition temperature of lanthanum. This was done on the basis strictly of the mechanism and there was no ω and no Green's function involved. Now, we come to the isotope effect. Again, on the basis of our pictures, not theory, strictly on our pictures, I predicted about four years ago that the isotope effect in uranium would be positive. Swihart and Garland, none of them even dared to predict anything; McMillan did - wrong, by the way. We finally did the experiment in Los Alamos - Fowler, Hill, et.al., and I - and, indeed, the isotope effect had changed its sign and had become positive. Now, if you look, we have 21 K. We have during the last ten years discovered superconductivity in essentially half of all the elements known. We have made more superconducting compounds, and not a single time was a theory of any help. How can you possibly say the theory has contributed greatly to the experiment?

It may have contributed greatly to some experiments with which I'm not particularly concerned, but in order to develop new materials, and that is what the technology will depend on, the theory has just given the wrong direction. Always, without a single exception. The materials used today were discovered by us twelve years ago, and the only new materials discovered since have been discovered by us and and we are not distinguished by being particularly bright or anything of the sort. We're the only group who doesn't believe that the theory has anything to contribute towards the transition temperature. Now this is a coincidence you can't possibly ignore.

Edelsack:

You wrote a very interesting article in a recent issue of Physics Today and I'd like to quote just one sentence from your article. You say, "Descriptions and explanations of superconductivity in the framework of the Bardeen, Cooper, Schrieffer theory is a beautiful approach". And then you go on and decry that this has gotten into the wrong hands. I'd like to ask you - it's a beautiful approach to what? You do claim that the BCS theory is a beautiful approach but it wasn't clear to me from your article - an approach to what?

Matthias:

Edgar, when you read sections from my article, would you please be so kind to read the whole section?

Edelsack:

But do you think that the BCS theory really has played an influential role in the development of experimental approaches to superconductivity?

Matthias:

In some instances, yes. None, which I've heard so far tonight. It has in the past given rise to new experiments and a certain amount of insight. But, in my opinion, it has long since fallen into the wrong hands. (Forgive me, I don't mean you.)

Schooley:

I hate to keep bringing up small points, but Dr. Hein and I both owe a few papers to suggestions by theorists in superconductivity that we undertake experiments on fairly specific materials. The reason I hate to bring it up is that it seems like nitpicking in the face of your statement. But, when you say "...without a single exception...", I feel that if I sit here dumb, so to speak,

I would be giving assent to this grand view, which really has a few holes in it.

Matthias:

Let me be quite specific. When I said not a single critical temperature has been raised on the basis of the theory, would you tell me where I was wrong?

Schooley:

Well, the semiconductors were examined as a class by you - I use it in the same editorial sense that you do - and the temperature was raised from zero to half a degree.

Matthias:

I'm sorry to disappoint you; we did not examine them.

Hein:

When we talk here we have to be careful. Dr. Matthias has an obsession with high T_c's. I mean there's a lot to superconductivity and the physics of superconductivity besides T_c. There's no doubt that BCS brought out many new facets. It's also true that in the context of above 18 K no theory has led to an increase in the transition temperature. The question is: To what do you attribute the three degrees increase between Nb_3Sn and Nb-Ge-Al? I don't feel that the fact that we went from 18 K to 21 K has increased my knowledge about how to increase T_c. Is this increase due to an increase in long range order, is it due to a more ideal stoichiometry, etc.? What do you feel the increase of three degrees is due to?

Matthias:

Just following our old beaten path a bit more closely than we did in the past.

Hein:

All right, now tell me what that is.

Matthias:

Well, look for the maximum transition temperature on the basis of the electron concentration and hope for the best.

Hein:

Yes, but with a three-component system, such as Nb-Ge-Al, I can change e/a by changing the relative concentration of Nb, Ge, or Al in the compound. I can wiggle e/a within lots of limits.

Matthias:

That's right. Sometimes it works, sometimes it doesn't.

Edelsack:

I wonder if we could get back to some of the other areas mentioned. As far as phenomena go, the area of fluctuations is one that has been opened jointly by experimentalists and theorists, but in large measure by theoretical ideas.

Scalapino:

I think these are areas in which the theorist has had an easier time because he's dealt with a system that is again more universal, in the law of corresponding states sense. One has been able to talk about a phenomenon and a calculation, rather than having to deal with specific materials. I think here there has been quite a bit of success.

Edelsack:

Dr. Webb, do you have a comment?

Webb:

Yes, I want to say something about the role of theory in our understanding fluctuations. As an experimentalist in the area, I'm a little reluctant perhaps, with the same attitude Dr. Matthias has, to say that the theorists have helped me. There has been a substantial amount of interplay between theory and experiment in the area. One asks the question: "Well, of a selection of theories, of which there may be half a dozen competing with each other to describe a phenomenon, which one comes nearest to being right"? I could say that what's happened is that the theorists have presented problems or opportunities to me, as an experimentalist, to try to sort things out. One finds, in fact, that new experimental results stimulate and help the theorist and vice versa. The leadership just switches back and forth between theorists and experimentalists. This is one of the features that's made this area so interesting in the last few years.

Shapiro:

That's quite true. It's very intriguing to see that the theories of fluctuations and noise and so on can fit various experi-

ments so well. But you've got to go very close to the transition
temperature in order to see this noise at all, and typically when
you're using a device, you're not working under such conditions.
In large measure, that is what I meant by my earlier statement
that the theories of noise have been of no particular use in terms
of making things work better.

Webb:

That must be true. I would also like to point out that the
original work of Schmidt and Ferrell certainly sparked all the
measurements on diamagnetic susceptibility and resistance right
above T_c. Of course Aslamazov and Larkin had previously published
their work on this problem in the Russian literature. However,
their method is very different from those now being used. Part
of the question is whether one is fitting a theory or not. I think
that one ought to ask whether the theory came first or not. There
are examples where the theory came first and the experiments later.

Edelsack:

Dr. Langenberg, did you have a question?

Langenberg:

I had a comment to make, Ed. I would back off from five
years to fourteen years, to include the BCS theory. Furthermore,
although I would admit to a predilection toward being suspicious
of theorists, we ought to remember what, I think, is probably the
major theoretical advance in superconductivity in the last nine
years. I mean Josephson's work. This was a purely theoretical
effort based, so far as I know, on no recognized prior experimental hints. I think every tunneler that I know who was in the game
before then had actually observed the dc Josephson effect, but had
attributed it to defective junctions. I suspect that if Josephson
had not done his calculations, the effect would have become sufficiently obtrusive so that some experimentalist would eventually
have "discovered" it. But nonetheless, in the whole area of Josephson effects in superconductors, the initial impetus was clearly
theoretical. In the work that's been done since in this area, I
think the theorists have provided a really essential framework for
the understanding of the extremely large number of phenomena that
one can see in Josephson devices. There are half a dozen different
kinds of these devices, and there are probably a hundred different
ways to make each kind. (One or two of them right and ninety-
eight of them wrong.) They're non-linear and they're active, and
the number of phenomena that you can observe if you simply make
one and play with it is incredibly large. You can see almost anything you want. But what is it you're looking at" Is there a

simple picture by which you can understand it? This is what the theorists have helped provide.

Edelsack:

During this first hour we've had a very interesting look at the science of superconductivity and after the break we'd like to resume and look at the technological and economic implications of superconductivity.

At a meeting last November on Naval applications of superconductivity, Dr. Langenberg used two terms, "megasuperconductivity" and "microsuperconductivity", to distinguish two broad but quite different classes of technological applications. In the megasuperconductivity class he included high power, big magnetic fields, and large volumes, as typified by such examples as magnets, machinery, and power transmission. In the class of microsuperconductivity, one is dealing with very small powers, magnetic fields, and volumes. Superconducting quantum devices, such as magnetometers and infrared and millimeter wave detectors, are examples of this latter class. These two classes bracket some thirty orders of magnitude in energy in which applications are being actively explored. As we talk this afternoon about the technologies, let us try to clearly distinguish which of these two classes we are talking about because I think they are significantly different. Let us start off by considering the "micro" class. Josephson tunneling, as was mentioned in the last hour, is an effect predicted and discovered within this last decade, but it's already forming the basis of a new technology. What I'd like to ask from you, gentlemen, is: will these Josephson junctions join the transistor and the laser as important new technologies during the next decade?

Webb:

Well, I certainly have no doubt that they will join these other illustrious devices as useful tools; they already have. Josephson junctions are providing us with measurement capabilities for magnetic fields, magnetic flux, and voltage or resistance that are many orders of magnitude more sensitive than we've had before. But I see no basis for thinking that there will be large scale utilization of that capability. There is, however, one other possibility for large scale utilization of the Josephson junction; that is, in computers. There have been several rounds of technological developments in the attempt to use superconductivity in computers. Several of the large computer manufacturers tried about five or ten years ago to use the cryotron, which is essentially a non-quantum device. And now there are some attempts, for the details of which I will refer you to Dr. Matisoo who is an expert in such things, to develop a superconducting memory device based on quantum superconductivity. These look very promising to me; the real advantage would be an enormous increase in speed. I believe that all of the elements that are required for cryogenic computers do now exist and, though it's an easy thing to say, all that's left are a few technical problems.

Matisoo:

Well, I think it's certainly true that there are a few technological problems remaining. In fact, let me try to put things into perspective a little bit. Essentially, I would say that any new technology which is not based on the current semiconductor technology, specifically silicon technology, is going to have a very hard time indeed. And that is, of course, because of the large investment which has been made in silicon technology. Nevertheless, I think it's certainly true that the Josephson junction is an extremely interesting device. It does have a chance, although a very slim one, because of it's very high performance capability. For example, logic gates have been operated at about 80 pico seconds logic delay, which is something like a factor of three faster than most semiconductor devices. And, on the basis of the rather good framework that we have for understanding what happens in Josephson junctions, provided by our friends, the theorists, one can extrapolate quite safely, I think, in many cases to even higher performance. Nevertheless, there is this big burden to be overcome of the already existing technology and the need for very low temperature operations.

Edelsack:

How about areas where there exists no technology, such as those utilizing the ac Josephson effect? Dr. Shapiro, what are your views?

Shapiro:

I think that in considering technology, the focus is no longer on what you will learn about the phenomenon, how exciting it is, and how interesting it is. Rather, the focus has to be put on what you are trying to do with this technology that you can't do with some other technology, and why the new technology is more desirable. It's only when you have answered these questions for yourself, at least tentatively, that you can go on to the next question: When? What are the constraints dictating when you need to have a system, a device in operation to do this particular task, whatever it is that you want to do? It's only when you've answered this question that you can begin to make a serious decision as to how to develop the new needed technology to do it. With respect to the ac Josephson effect, I think we've seen a very logical development. The first really significant utilization of the ac Josephson effect, I believe, has been in the e/h measurements of Dr. Langenberg and his colleagues and the use of these measurements in voltage standards. And the reason that this technology was selected for this purpose is that it is unique and represents an order of magnitude advance over prior technology. It seems to me that the realization of the uniqueness of the ac Josephson effect technology in this area has really been the spur for its development. I hope that we don't have to rely in the future, in such areas as far infrared detection, for instance, entirely on this uniqueness criterion as a spur for development.

Langenberg:

I'd like to make a comment on the area of "microsuperconductivity". I think one of the things which must be done is to encourage broader use of Josephson device systems by getting them out of physics laboratories (where they already are flourishing, as Dr. Webb has pointed out) into chemistry and biology laboratories and maybe even out in the field. We must tell people in these areas a little more clearly and simply and a little more intensively than we already have just what can be done with these devices.

Scalapino:

One of the things that's going to make these devices more useful will be non-superconducting electronics that operate at cryogenic temperatures.

Webb:

Perhaps we should follow up with Dr. Shapiro's original suggestion, and talk for a couple of minutes about the unique properties that can be obtained with quantum superconductivity.

Edelsack:

Very good.

Webb:

If you're willing, I will start it off.

Let's talk about what one can do with the flux detector, which is more or less a slow speed device and which can be used as a flux meter, or a magnetometer, or as a device for measuring magnetic susceptibility, or as a voltmeter or ohmmeter. In fact, we have one device that can measure 10^{-15} to 10^{-16} volts. That's a factor of 10 million times more sensitive than is available with any room temperature instrument, such as a standard nanovoltmeter. It is available now to anyone who wants to take the trouble to put it together. This level of sensitivity is of course way below the level of noise that one has in essentially any room temperature circuit. So this kind of measurement is restricted to low temperature phenomena. The power which can be measured is as low as 10^{-23} watts, which is a power level that one would associate with thermal noise at the temperature of liquid helium. These are capabilities that exist now, and there are even alternative schemes to do these things. In the magnetic measurements area, one can measure, on a more or less routine basis, magnetic field intensities between 10^{-9} and 10^{-10} gauss; that's something less than a billionth of the magnetic field of the earth.

Someone raised the question a minute ago, "Can one make this kind of measurement in a room temperature environment?" I would cite one application which, in fact, was done at room temperature with a sensitivity of 10^{-7} to 10^{-8} gauss. Ed Edelsack, our Chairman, David Cohen, and Jim Zimmerman at the National Magnet Laboratory used superconducting magnetometers for a measurement of magnetic fields associated with the beating of the human heart. This kind of measurement gives information about the operation of the heart which is not available in an electrocardiogram. So here we have clear evidence for the use of this kind of instrument to do something which you cannot do in any other way, and it's at room temperature. I'm sure that more of these applications will come along. One of the applications that's already approaching routine practice is the use of superconducting magnetometers to measure the magnetic susceptibility of rock.

Edelsack:

Dr. Finnegan, we heard earlier in the week, in your lecture, about the possibility of converting over from the chemical voltage standard to a voltage standard depending on the Josephson effect. Do you see this as a real possibility in the near future?

Finnegan:

Yes, I think not only is there a real possibility but that it is only a matter of time before the National Bureau of Standards here in the United States and most other major national standards laboratories will be establishing their unit of voltage in terms of the Josephson effect. At NBS it is primarily a matter of internal bookkeeping to convert from a chemical unit to a Josephson effect unit, once the latter is truly operational. At present it is fair to say we are in a necessary transition period during which both the chemical volt and the Josephson-effect volt are being maintained side-by-side. Within a matter of months (at most a year), the U.S. national voltage standard will be based on the Josephson effect.

One of the big problems faced by most of the national standards laboratories has been the fabrication of suitable Josephson devices, particularly of the 10 mV tunnel junction variety. These devices are usually a bit tricky to make with just the right characteristics and generally deteriorate rapidly at room temperature. An important future development will be the fabrication of durable 10 mV devices which can survive lengthy room temperature storage. Tunnel junctions (with niobium oxide barriers) have been fabricated, for example, by Schwidtal and co-workers at Fort Monmouth. These have been successfully stored on the shelf for relatively long periods of time at room temperature. The development of suitable and reliable off-the-shelf tunneling devices is an important area of future research and a necessary one if a voltage standard based on the ac Josephson effect is to enjoy widespread use.

I would also like to mention briefly some other applications of superconductivity which will play a role in the continued development of Josephson effect voltage standards. In particular, there is the cryogenic voltage divider which can share the helium dewar with the junction device. The use of superconducting leads and switches for resistor connections in the divider gets rid of two of the major problems in building accurate voltage measuring instruments at room temperature. The use of a superconducting null detector, no longer limited by the thermal noise at room temperature, will permit better resolution and higher accuracy. At present, using room temperature voltage measuring techniques, one can establish a volt about an order of magnitude better than with conventionally used standard cells. One has an immediate possibility of yet another order of magnitude beyond that with superconducting instruments.

Hein:

May I ask a question? There's no question that Josephson devices will be useful as a laboratory tool. But I thought the question was - "in technology". Now we've heard about computers and that seems to be one possible application. How long is it going to be before someone comes up with a superconducting device which will be used in everyday technology? Aside from their use as a laboratory tool to establish fundamental constants or to measure magnetic susceptibilities, and things like this, where do we foresee Josephson devices coming into our technology?

Hamilton:

We've got a long way to go in our support technology. This support technology can be completely separated from the technology of the superconductor. As anyone who's ever bought a superconducting magnet knows, you'll spend your entire budget in helium to keep it cold indefinitely. Dewars are just now beginning to be made which enable us to work continuously at low temperatures. In the past, you've had to make them yourself, which has inhibited the whole field. I think that we're looking at just the beginning of the time when these low temperature techniques, and I extend this to more than just superconductivity, can begin to move out of the laboratory. This technology is moving out not only because of the sensitive devices, although that is very essential, but also because all of the other support technology is being built up. We have helium readily available, even in Louisiana.

Edelsack:

That's a very good point that you bring out, Dr. Hamilton, that some of the techniques are moving out of the laboratory into the realm of practicality. This leads me to a question I'd like to ask you gentlemen. Do you think there exist a sufficient number of engineers and scientists with training to carry forth these developments, which to date have been largely done by physicists in the laboratory? Can someone who so desires be trained in this area?

Hamilton:

I can give you just one example from our own experience. One of the experiments that we're doing at Louisiana State is associated with the development of a cryogenic detector for gravitational radiation. It's somewhat larger than the usual low temperature experiment. We have a mechanical engineer at Louisiana State who, once he was convinced that the cryogenic technology was nothing very mysterious, was able to move into the field very easily. He did an outstanding job on the design of a rather large cryogenic experiment. I should say that the experiment has not been built yet - it looks like an outstanding job on paper.

Edelsack:

Your impression is that one with good technical background can move into this new area of technology?

Hamilton:

Yes. I'd say that we have to. The people who are working in the field have to let the mechanical engineers know what it is that they need.

Stekly:

I'd like to make a few comments based on my own experience. I have always had, in most of the organizations that I have been at that were non-government, an excess of qualified people that I could put to work on any of the required problems. I think this is the case today. So, I think the moving out of the laboratory into the real world will certainly not be hindered by the availability of qualified people.

Edelsack:

One of the problems which arises in using, say Josephson devices, is reproducibility. In the laboratory one can build one of these or ten of them, but when you really talk about the technology, you want to be able to build hundreds of them of which the major portion work. This requires subtle techniques. But, do you think such techniques can be carried over from existing technology into superconducting devices?

Langenberg:

I might make one comment about that. I have, in the last week or so, had occasion to look through a good microscope at an integrated circuit, one which was bought for $1.25 I think. It had 53 transistors and God knows what all else on it. It was the first one I had actually seen "in the flesh", rather than in a picture or on a magazine cover. It was most impressive! If that can be done for $1.25, I claim that any Josephson junction or collection of Josephson junctions can be made with equal or greater facility.

Edelsack:

Well, that's very hard to believe. Would they be as stable? Would they be recyclable?

Langenberg:

Well, that depends upon what kind of Josephson junction you're talking about. Schroen at Texas Instruments and Schwidtal at Monmouth have made storable, reproducible tunnel junctions. And they work! Most of us who have used tunnel junctions haven't really had to do this. A high level of junction yield and reproducibility hasn't been a major problem in experiments, and so we haven't put a lot of effort into achieving it. But it can be done. People who have wanted to make point contacts that don't have to be fiddled with every time you turn around or put your foot down hard on the floor have been able to do that too. Weak link structures of various kinds are intrinsically stable so long as you don't blast them with a big voltage. So, if the need arises to have something which you can turn on in the morning and turn off at night which will last, I think it can be done.

Gregory:

Dr. Peter Mason at JPL was very much worried a couple years back about the yield problem. The manufacturers have a rather high criteria. They have to have a very high yield before they consider that the time involved is worth it. Then, Mason didn't feel that it was. Do you think this has changed?

Edelsack:

I'd like to ask you gentlemen a question in terms of devices. Suppose we could come up with a superconducting material which has a transition temperature of 25 K or so. Would this significantly change the potential of superconducting Josephson device technology and microsuperconductivity in general? Suppose we have materials with transition temperatures in the range of 25 K to 30 K. Would that significantly change the field? Would it be a major factor, would it be a major breakthrough if we had such materials?

Webb:

Ed, I'd like to back up a little bit in my answer and try to speak to a subject that Bob Hein raised and that's relevant to the next subject, which would be the stability of devices, and to this one too. Hein asked about the technological applications of these things and what it will take to bring them out of the laboratory. We might ask at the same time what kind of motive there is going to be to do the development effort that's necessary to get the reliability that one can now get in integrated circuits. It seems to me this is an extremely important question, because what we've got here is a capability which exceeds the capability we've had before. It doesn't let us do better, except with the possible case of the computer, what we could do before. If you want to measure one volt faster or something like that, you might as well do it the old way. If you want to get much more sensitivity or much higher precision, operate at lower signal levels. That's the kind of thing you can do with these devices. So what I would say is that what we're getting is a new range of capability for which technology needs to find uses before these things will turn up as widely used devices.

Edelsack:

Would you then class superconducting devices as one used to classify lasers, as a solution seeking a problem? Do you think that what we're really looking for are problems which superconducting devices can be useful in solving?

Webb:

That sounds like a good analogy to me. When lasers first came out, they rapidly found a good many applications. One might say the same here. Nevertheless, it is difficult to predict how the technology will grow, because the applications are likely to be new.

Langenberg:

Another comment with respect to the question of routinely usable devices. Maybe I shouldn't say "devices", but "systems", because that's my point. The device that has fascinated many of us here, the Josephson device, is very small and very simple. In any system that one might use in the field or the non-physics laboratory, it is an absolutely minuscule part of the overall system. If you are going to have chemists or geologists using these things, you're going to have to sell them the whole system, not just a superconducting device. That's a problem that none of us have really looked at or faced. It is, I think, a rather considerable engineering problem which must be solved before these things are going to be in common use for almost any application outside the physics laboratory. I don't know whether it's useful to draw analogies in this case, but I might compare the crude NMR systems which were built by my graduate student colleagues in their own labs with what Varian now sells with a superconducting magnet and a 250 MHZ rf system and all sorts of sophisticated circuitry. It's a big beautiful box and they sell them to chemists who have lots and lots of money. Chemists can flip on the switches and the system works. That's the kind of thing which must be done with microsuperconductivity devices based on Josephson junctions if they are really going to see wide use.

Rose:

Let me perhaps lay a few doubts to rest on the subject of yield and reliability. About five years ago Margaret MacVicar and I were doing tunneling experiments on niobium single crystals. We were not looking for Josephson tunneling at that time, but it was just as easy to get Josephson as non-Josephson tunneling. After we acquired experience at it we produced about 100 junctions, 85 of which were good. We found that typically you could take one of these junctions out of the holder after you've made your measurements, disconnect all the connections, put it in the drawer, and let it sit in glorious Cambridge air for about six months. You could then take it out of the drawer, reconnect it, make your measurements, and get the same characteristics. So as far as durability, reliability, and yield, I see no problem. It's simply that niobium oxide is a very durable, reliable barrier, as long as you know how to make it.

Shapiro:

I'd like to come back to the question, which was originally asked a few minutes ago, of how a superconductor with a transition temperature of 25 K would influence the technology of microsuperconductivity. I'd like to try to follow that up, not so much to actually answer the question, but to discuss it. I'll make one assumption: that this superconductor with a transition temperature of 25 K has a correspondingly high energy gap. If it does, the frequency response to applied radiation will extend to around 10 microns in wavelength. Then it is a potential candidate for use with the CO_2 laser system. If it is a potential candidate for use with the CO_2 laser system, what do you want to use that system for? If you want to use that system to transmit telephone communications, then what does it buy you over some other kind of system detector? Well, you say, it buys you sensitivity. All right, how can you use that sensitivity? It seems to me there are only two ways. One is that you need fewer repeater stations, you can put your detectors farther apart. But if you do that, then perhaps you are worried about the real earth - the earth bends, it has mountains, it has bumps, it has holes. So maybe economically, from an overall point of view, there isn't any advantage to be gained by putting your repeater stations further apart. The only other consideration is that if you're transmitting this laser radiation through a pipe, maybe because of the increased sensitivity you can afford much more loss in this pipe and, if so, then you don't have to evacuate the pipe. It seems to me that considerations such as these are ultimately going to determine the usefulness of any of these devices that we ourselves create in our laboratories.

Edelsack:

Dr. Hein, you had a question.

Hein:

Let me come back to this one. Why talk about the 25 K superconductor that may someday exist? How about using the 21 K one that does exist? Isn't there a well-defined energy gap in Nb-Ge-Al?

Matthias:

Is there an energy gap in Nb-Ge-Al? Sure. Unless you try to get rid of it.

Hein:

How big is it?

Matthias:

I would guess $3.5 k_B T_c$.

Hein:

Is that a theoretical prediction?

Edelsack:

Gentlemen, at this point let us look at the "macro" side of the house. And here again, I'd like to ask if 25 K transition temperature material would result in significant benefits in the development of superconducting motors, magnets, and power transmission lines. Dr. Smith, would you care to comment?

Smith:

Well, I'll try. The first thing I would like to say is that we are just getting a good start on developing the potential of the superconductors which are available today. Let us finish that job. Now back to the question of what a higher transition temperature would do. Well, to answer this question, I think you have to look at those characteristics of the superconductors available today that allow us the projected potential in superconducting electric machines. The characteristic that gives us this potential is that with superconductors we can produce higher magnetic fields than have been economical before by using a winding carrying very high current density. It appears that the possibility of using currently available superconductors to produce higher magnetic fields at higher current densities offsets the penalty of a helium temperature cryogenic refrigeration system. Of course, if a 25 K superconductor would give us comparable magnetic fields with comparable current densities, then, undoubtedly, a less complex cryogenic refrigeration system would be required. However, I suspect that one would have to go into the tradeoff here of the optimum economic temperature; as we go down in temperature, in general, we have higher magnetic field capabilities and higher current density capabilities. So, really, I think you could give us an additional range of freedom in the organization of the operating conditions for superconductors in electric machines.

Matthias:

I'd like to make one short comment. A T_c of 18 K will give you 200 kilogauss, in one way or another at $4°K$. A 21 K T_c, according to Foner and the rest of the group at the National Magnet Laboratory, will give you 400 kilogauss. So it's quite clear that, as it has in the past, an increase in T_c of three degrees yields a doubling of H_{c2}. I remember ten years ago when I was assured by the plasma physicists that controlled fusion would be possible if a magnetic field of 67 kilogauss over an arbitrarily large volume were available. Well, controlled fusion has not been achieved yet, but maybe it will be with 670 kilogauss.

Edelsack:

Mr. Haid, from the transmission side of the house, what are your feelings? Are the present materials a limitation?

Haid:

I don't think the present materials are a limitation, as such. We would always want to look carefully, though, at what higher transition temperature materials would buy us, because at first thought you would think they might buy a great deal. But we do have to use the same line of reasoning that both Drs. Shapiro and Smith employ; that is, if we do have a high transition temperature material, let's look at its ac losses. I mention this because, I think it's generally recognized, the first big pinch as far as underground power transmission is going to occur in about 1985 with ac power transmission over relatively long distances - five to fifty miles. I don't mean to imply that there are no arguments for superconducting dc power transmission. It's just that I think the pinch there is going to come later. We should address ourselves first to the ac problem. Looking at the ac losses helps us decide whether the higher transition temperature is really going to be exploitable. If the ac losses are low, then I would next look at whether we can fabricate this material and the cost of fabrication. Most of the studies today have indicated that cost of the superconductor itself, installed in the line, is appreciable, as is the refrigeration. So we're going to have a tradeoff here of the cost of the superconductor against the refrigeration. I would sum it up by saying that if someone discovered a material with a very high H_{c1}, as well as a high transition temperature, which could be fabricated with reasonable facility, it would undoubtedly have a very large impact upon ac superconducting power transmission. Even before that, if someone learns how to lay down Nb_3Sn, or any of the other high transition temperature materials, in such a way as to obtain lower ac losses than have been evidenced to date, this itself would have a very large impact. This is something that we (Union Carbide) intend to study. We are fortunate, those of us who are interested in superconducting power transmission, in that we do have a fairly good lead time. No one is talking about installing such lines before 1985. But we do realize we also have to start working now if we're going to have anything to install in 1985.

Edelsack:

How about the magnet side of the house, Dr. Stekly?

Stekly:

Well, let me make several comments. I believe it was in 1960 that I picked up the New York Times and read that Nb_3Sn remained superconducting to 88 kilogauss. That statement in the paper took approximately one day to change a magnet program for a cryogenic sodium magnet that was almost in the hardware stage into one looking at Nb_3Sn for its potential. Now, if tomorrow I picked up the New York Times and Dr. Matthias had discovered a 25 K material, I would say the group has done a good job and this is what we expect of them. However, I wouldn't change any of the programs with which I'm currently involved. I believe that in the long run 25 K is more desirable than 18 K, which is more desirable than the 9.5 K for Nb-Ti, of which I make most of the magnets for devices, such as rotating machinery and energy storage, or plain dc magnets. Nevertheless, I don't wish to discourage anybody from going ahead with efforts to find a 25 K material.

Matthias:

I just would like to ask another question. If, however, you read in the New York Times that the material at 25 K was highly ductile, what would you do?

Stekly:

That is a very important fact.

PANEL DISCUSSION

Matthias:

Then would you change?

Stekly:

Yes, I think it would be very close to the kind of impact that Nb_3Sn had back in 1961. We do not make a large majority of our devices out of the 18 K material, for mechanical reasons alone. It's ductility has lead to widespread use of Nb-Ti. I think ductility would lead to the widespread use of any higher critical temperature materials.

Rose:

I guess of the assortment of talents up here, I'm the man with the hammer in his hand. As a consequence, I think it's time to speak up because this is the problem: metallurgists know how to make things, and generally when they do exist, the materials problems always seem to come last. I can remember that in 1946, everyone thought that there would be nuclear reactors in every kitchen. But, among other things, there were a few materials problems. Nothing we couldn't slap a patch on, you understand. As of today, many of those "trivial" materials problems remain to be solved. The point is that motor and magnet building, right now, depend on the ability to fabricate very sophisticated composites of Nb-Ti alloys in Cu without breaking them up. The first such composites were made, I think, at IBM and in my laboratory, and I'm still very impressed when they don't break up, when they actually work. Using the A-15 compounds, even under dc conditions, depends on fabrication of a cheap, reliable, fine conductor, stabilized against magnetic instability. Without that, even a large dc machine operating at 15 K has to be a dream. We're only now beginning to tackle the creation of low loss ac materials. But, as usual, I think we're going to find at this juncture that the materials are going to be the chief bottleneck.

Edelsack:

Dr. McAshan, at your laboratory in Stanford you're putting together a 500 foot superconducting linear electron accelerator for which, I believe, the cavities have been made of Nb. If some morning in Palo Alto you picked up the paper and saw that some group in the country had been fabricating materials with a T_c of 25 K, do you think that would significantly change future superconducting accelerator technology?

McAshan:

It certainly might change the one we're building, because we haven't built all that much of it. Once again, the stuff certainly has to be ductile, because we form it into a tube with many wrinkles in it. The second thing is that it has to have a high H_{c1}. I really don't know what H_{c1} in a 25 K material would be like.

Matthias:

What is your H_{c1} now?

McAshan:

For pure Nb?

Matthias:

For the average cavity.

McAshan:

I would prefer not to say the average cavity. Let me talk about the best cavity, in which case we would say 1200 gauss. For the worst, H_{c1} is more of less negligible. Average would be around 200 gauss I would guess, but that of course is our problem.

Matthias:

Why should the H_{c1} be so low? After all, according to Abrikosov it should be much higher than that.

MacAshan:

The problems of H_{c1} in pure Nb are that we don't know the right words to say over it. One has to know the proper incantation to make it work out.

Matthias:

Why should the H_{c1} be so far below the theoretical value? You know the kappa of Nb and all such things. The H_{c1} should be much higher. Why is it so low?

MacAshan:

I don't know if one should associate the breakdown fields

PANEL DISCUSSION

that occur in these cavities with H_{c1}. Possibly they should be associated with the field at an asphericity on the surface. These cavities were annealed mostly in ultrahigh vacuum. When you anneal them, you get thermal faceting, pyramids, if you will. And so the field is concentrated at the point of the pyramid.

Shapiro:

If you want to pursue this particular line, I can refer to a paper by J. Halbritter in the January, 1971, issue of Journal of Applied Physics (J. Appl. Phys. 42, 82, 1971). This is the issue in which the last Applied Superconductivity Conference proceedings were published. He had an intriguing idea - I don't know whether anyone has tried to prove it out in detail - that the breakdown fields were associated with phonon generation in cracks. The high local r.f. fields excited phonons. This loss mechanism would come into play at a relatively low field compared to the theoretical H_{c1}. So there's at least another mechanism that's possible, again associated with the structure, the detailed structure of the materials, whether pyramid or valley.

McAshan:

Well, let me give you another one, if you like. If you pull the niobium down too slowly in ultrahigh vacuum, you'll form a niobium monoxide in clusters.

Matthias:

You should check the pressure.

McAshan:

Yes. The equilibrium partial pressure of oxygen which saturated that 2% oxygen and niobium solid solution is approximately $(4)10^{-11}$ at 1700 K.

Matthias:

You can do better than that.

Edelsack:

Gentlemen, let us proceed now to discuss some other facets.

Webb:

At the risk of saying the obvious, it seems to me there's a moral to this discussion. That is, the critical temperature is not the only important property of the superconductor.

Edelsack:

Let's talk about magnets for a moment. I would like to ask one of our guests here from Germany, Dr. Martinelli, for his comments.

Martinelli:

Let me come back to Dr. Mathias's comment on the superconducting magnet requirements for plasma physicists. In fact, these requirements are becoming more and more exacting. The tendency is to discount the possibility of a future fusion reactor being realized with fields lower than 100 kilogauss. In fact most people working in this area feel that the required fields will be much greater than 100 kilogauss, and therefore, commercial production of very high field superconducting materials such as $Nb(Al_{0.8}Ge_{0.2})$ or Nb_3Al is important. There is not much concern about the critical current of these superconductors; the large volume of reactors results in the required current densities being rather low, probably of the order of $20 - 30$ A/mm^2.

Such reactors would be feasible when future materials having critical fields of 300-400 kilogauss become commercially available at a price allowing economic application in reactors.

I would also like to comment about the use of superconductors in plasma experiments. Almost all plasma laboratories are engaged in projects which involve superconducting magnets. Here too, the trend is toward larger volumes and higher magnetic fields. We plan at Garching a superconducting stellarator with a volume of about 10 cubic meters, a stored energy of about 60 megajoules, and a field of 40 kilogauss in the centre and of 65 kilogauss at the windings. But our physicists would prefer even larger volumes and fields.

Edelsack:

There is a question from the audience.

Questioner in the Audience:

We've heard about the greater advantages which will result from the large scale application of superconducting technology. But, has any thought been given to new disasters which might result?

PANEL DISCUSSION

Haid:

In the case of superconducting power transmission, this consideration is often brought up and we have given it much thought. When you establish a cost for what you're proposing, you have to take into account reliability. It depends upon the application. When you're talking in terms of transmitting more than 1,000 Mva power from the Indian Point nuclear complex on the Hudson River down into Manhattan, reliability is of the utmost importance. One of our approaches to that problem is to consider double circuits. If someone putting in a new sewer line takes his backhoe and runs through your power line, what's Manhattan going to do without you? We have established the cost of double circuit lines, where you separate lines far enough so that they are for all intents and purposes independent. You've got to be a little bit careful that you don't restrict yourself to this approach since, particularly in the Northeast, there are other possibilities. I'm sure that Indian Point will be connected to New York via more than one circuit. You can go via the back road, so to speak, because lines extend out from Indian Point in other directions. So you may not have to have 100% redundancy in the form of a second circuit that can carry full power. I would say that yes, in the case of power transmission, one has to consider this (the prevention of a disaster). It's an obvious question, it is being considered, and it doesn't look too bad.

Edelsack:

Anyone else have any comments on that?

Coffey:

The primary application of superconducting technology that we're working with at the Stanford Research Institute right now is magnetic levitation of vehicles. Since these vehicles will carry people, safety becomes a prime consideration. If it isn't safe, you're not going to build a system, however beautiful it might look. This refers to the question that was asked earlier as to whether a 25K superconductor would help. It really wouldn't help a whole lot. There are a lot of things that can be done with the present technology that have not been done. We don't need a higher transition temperature to build a magnetically levitated vehicle. The question of ac or inherent stabilization was brought up earlier. In the case of the train, I feel that at the moment ac stabilization is not as reliable as the Kantrowitz and Stekly cryogenic stabilization that was worked out a few years ago. Where you're going to carry passengers you do not necessarily use the most exotic techniques that are available, you use the safest ones. And the same sort of thing would apply to transmission lines, and motors, and generators that are built where people are around.

Edelsack:

In the remaining few minutes I'd like to get ideas from you gentlemen on the following area. A few weeks ago Secretary of Commerce Maurice Stans, in Congressional testimony, stated that "The relative strength of U.S. technology in the world is on the decline." Do you think that the U.S. is losing technological leadership to other nations in the field of superconductivity?

Langenberg:

Well, in the area of microsuperconductivity, I have a feeling that most of the work on laboratory applications of devices based on Josephson junctions has, in fact, been done in this country. There has been little extension of this work to in-the-field systems, so I think it's difficult to say what might happen. But there's obviously a great deal of interest abroad. I hesitate to say that more is being done abroad in the area of megasuperconductivity, because such a statement implies the people here aren't doing enough, or they don't have enough help, or something, which may be true. But I do have the feeling that in both microsuperconductivity and megasuperconductivity the science and a little bit of the beginning technology is well advanced here. There is perhaps not as much actual conversion of this into working hardware so people can get used to using it and see what the actual operating conditions and economics are as there should be; I have the feeling that more of this is going on abroad.

Edelsack:

Dr. Matthias, what was your comment?

Matthias:

I wanted to comment on an article in Soviet Life two years ago, in which the Russians published the fact that they were building a superconducting Nb_3Sn cable. They said, however, that this was only a preliminary arrangement since their theorists had assured them of a 40 K superconductor within two years. So I think the Russians aren't any better off than we are.

Webb:

I wanted to second everything Langenberg had to say about the activity in this field in the rest of the world as compared to the United States. I've just come back from a trip which allowed me to see some of the work that's going on. It seemed to me that in Great Britain the technological applications of macrosuperconductivity have been carried a lot further than in this country. I have the feeling that we are really slipping in our usual role as a technological leader.

Edelsack:

Dr. Hein, you spent a year looking at the field abroad. Do you share the views that somehow efforts in Europe are proceeding along at a greater rate than they are here in the States, at least on the macro side of the house?

Hein:

It's difficult to say, because England had a shot in the arm; they had a National Ministry of Technology. But that seems to be drying up. The only thing I can say is that in Europe there are people building transmission lines, real ones that you can walk on and jump on, while people in the States are still doing paper studies. They have motors which work over there, while we're just starting. We have some that work here. I don't know whether we can say they are ahead, because if you give an American industrial concern a million dollars, they can build you a motor.

Edelsack:

But at the present time, obviously the potential exists there. Would you say that we are not up to speed with the Europeans or with the U.K. effort in machinery, or do you feel that we are about at the same point?

Hein:

You have to say the U.K.'s ahead because they have the biggest motor working. I mean they had it working or it was working. They have a transmission line which is "working". They are doing a lot of work on dielectric breakdown with helium. They have a 50 meter length of cable. The only reason they stopped at 50 meters was that was all they could line up at the University where the project is being done and transport in a truck. They have advocated the use of helicopters. They can run the cable up to 250 meters, fly it out, and lay it right down in the trench. So, I would say that's a measure of being ahead. I don't know the secrets of the various industrial companies. But on the surface of it, yes, I would say they are ahead.

Edelsack:

Gentlemen, before we close does anyone have any particular comments:

Webb:

I would like to make a comment. Someone asked a few minutes ago whether we were going to be introducing any new hazards with superconductivity. I'd like to look at that problem from a slightly different point of view. We look at some of the hazards that our country is facing at this point, we see some of them associated with an old and somewhat sleepy industry - the electric utilities. And we see some hazards associated with that sleepiness. We can see in superconductivity a possible solution to some of the problems of that industry. It seems to me that it behooves us to try to use this new kind of science to develop technology that might solve some of these problems.

Edelsack:

By way of conclusion, let me say that in tomorrow's sessions we will have two summary discussions by Dr. Bostock and Dr. Langenberg. But, it appears that superconductivity is alive and well; it's had a remarkably exciting past and our discussion today certainly gives hope that it will have an equally exciting future. I want to thank the audience and the panel for participating in the discussion.

Part V
Conclusions and Summary

THE TECHNOLOGICAL IMPLICATIONS OF SUPERCONDUCTIVITY IN THE NEXT DECADE

D. N. Langenberg

Department of Physics and Laboratory for Research on the Structure of Matter, University of Pennsylvania, Philadelphia, Pennsylvania 19104

and

R.A.I., Inc., Newport Beach, California 92660

"Why does this magnificent applied science, which saves work and makes life easier, bring us so little happiness? The simple answer runs:--because we have not yet learned to make a sensible use of it........."
—A. Einstein, Address to the Student Body, California Institute of Technology, 2/16/31.

Before I address myself to the questions implied by the rather awe-inspiring title the organizers of this course have assigned me, I would like to issue a disclaimer. I am not now and never have been a card-carrying prophet or seer. By saying that, I hope to forestall any future reminders that I said something perfectly ridiculous here. (Of course, I reserve the right to call attention to my farsightedness in case anything I say actually turns out to be correct.) The actual development of superconducting technology during the next decade will be determined as much by unpredictable factors like the health of the national economy and the actions of the planners who operate within it, or the possible appearance of another Brian Josephson, or a room temperature superconductor, as by the scientific and technological knowledge we have today. Nevertheless, one can discern some interesting developments and possibilities. I would like to discuss these in terms of the following questions: (1) What knowledge and demonstrated capabilities in the field of superconductivity do we now have? There are some pertinent collateral questions here, such as: Who are "we"? Where are "we"? Who has supported our work? (2) What can we do with superconductivity within the next decade, given sufficient effort and support? (3) What should we do with it? We might also ask what we will do with it, but answering that question would require more courage and a better crystal ball than I possess.

Let us set the scene for our discussion by considering the collateral questions mentioned above. In 1966, Schmitt and Morrison[1] made a rather thorough study of money and manpower in the field of superconductivity. They estimated that nearly 150 million dollars had been spent on research in superconductivity since the discovery of the phenomenon in 1911, two thirds of it in the United States. They documented a spectacular growth of interest in the field in the late fifties and early sixties. In 1964 in the U. S., about 375 Ph.D. level scientists were active in the field, supported at a level of approximately 18 million dollars per year. Roughly 50 per cent of this money and manpower was being expended in universities, a little more than 40 per cent in industrial organizations, and a little less than 10 per cent in government laboratories. The Federal government provided about 60 per cent of the financial support, industry about 25 per cent, and the universities about 15 per cent.

What is the present picture? I have not attempted a detailed study like that of Schmitt and Morrison, but some indication may perhaps be found in the data shown in Fig. 1. One curve shows the number of articles on superconductivity published in Physical Review Letters during the six years up to and including the present one. The other shows the number of publications and patents referenced in Superconducting Devices and Materials.[2] Publication in Physical Review Letters is supposed to mean that the work is an important and timely contribution to an active field of basic research, hence worthy of rapid publication. The rate of publication in Physical Review Letters of papers in a given field should therefore give a fairly current indication of the level of interest in basic research in that field. Superconducting Devices and Materials is a comprehensive survey of the (open) world literature on its subject, including basic theory and experiment, and should give an indication of the worldwide interest in superconducting science and technology. It is clear that the trend in basic research in superconductivity is down, while the trend in basic and applied superconductivity is up! It is difficult to infer anything about money and manpower expenditures from these data, but it may be noted that if $25,000 - $30,000 is taken as a reasonable estimate of the average current expenditure per publication[1], the total 1971 worldwide expenditure comes out at 25 to 30 million dollars. This suggests that the overall financial support picture has not changed

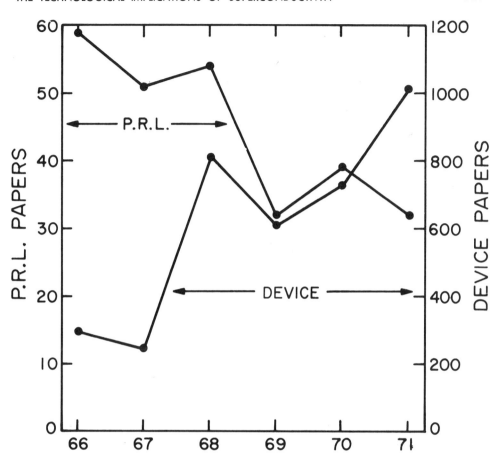

Fig. 1. Superconductivity publication rates. The curve labeled P.R.L. (left scale) results from a count of papers on superconductivity in Physical Review Letters; these data should be quite accurate. The curve labeled DEVICE (right scale) results from a count of "first author" notations in the author index of Superconducting Devices and Materials[2]; these data may differ from the actual numbers of distinct publications and patents by 5% to 10%. The numbers for 1971 were obtained by doubling the counts for the first half of the year.

greatly since 1966[1], while there has been a shift of interest from basic toward applied superconductivity.

The several hundred million dollars which have been invested in research on superconductivity have brought us an understanding of a complex, almost magical natural phenomenon. Have they also brought us the basis for an important technology? This is an all-important question I would like the reader to keep in mind in what follows.

What is the present state of our knowledge of superconductivity? The answer to this question can be found in the other lectures in this volume. Let me try to summarize the basic points.[3] First of all, we have a microscopic theory of superconductivity, the BCS theory, in which a phonon-mediated attractive interaction leads to pairing of electrons of opposite spin and momentum. The relatively large size of the resulting pairs (the coherence length) and the consequent strong overlapping of pairs then leads to the formation of a phase-coherent macroscopic many-body quantum state split off below a continuum of single particle states by an energy gap. This state is the superconducting state. Because this single quantum state is occupied by a macroscopic number of particles, the amplitude and phase of the wave function are observable in the same sense as they are for a classical electromagnetic field. The BCS theory together with its subsequent elaborations appears to give a quantitative description of almost all superconductor phenomena. Like many successful microscopic theories, the BCS theory was preceded by phenomenological theories which remain useful for the treatment of many problems of superconductivity. These include the Ginzburg-Landau theory, which has been developed into an extremely useful tool for the study of situations in which the space and/or time dependence of the superconducting wave function or order parameter is important. The response of a superconductor to electromagnetic fields is inherently nonlocal, and this feature is incorporated into a theory developed by Pippard. However, many situations can be adequately described by a local theory due to the London brothers together with the two-fluid model of Gorter and Casimir. Recently there has been considerable progress in the theory of fluctuations in the superconducting state.

These theories successfully account for the principal observed features of the superconducting state. These include, first of all, zero electrical resistance and essentially perfect diamagnetism (the Meissner effect). Then there is the existence of the two major types of superconductors, Type I and Type II. In the first, the superconducting state is thermodynamically stable at temperatures below the transition temperature and magnetic fields below the so called thermodynamic critical field. Transition temperatures and critical fields for bulk samples of this class of superconductors generally lie below 8 K and 1 kG respectively. For appropriately simple sample geometries, magnetic flux is completely excluded from the superconductor at fields below the critical field. In Type II superconductors, a second state, the mixed state, is stable in magnetic fields ranging from a lower critical field (less than the thermodynamic critical field) to an upper critical field (greater than the thermodynamic critical field). In this state, flux penetrates the material in the form of lines or bundles of flux quantized in units of the flux quantum, $\Phi_o = hc/2e \cong 2 \times 10^{-7}$ G cm^2. The cores of these fluxoids are normal, but the remainder of the material remains superconducting. Electrical conduction at zero frequency is lossless at least at low current densities, but the motion of the fluxoids in alternating fields leads to non-zero ac losses. Because the transition temperatures and upper critical fields of Type II superconductors range up to 21 K and several hundred kilogauss, these materials are of particular interest for applications involving large magnetic fields and high currents.

The quantization of the fluxoids which exist in the mixed state of Type II superconductors is a consequence of the macroscopic phase coherence of the superconducting state, together with the requirement that the wave function describing the state be single valued. This also leads to the quantization of flux in closed superconducting loops. The phase coherence of the superconducting state took on new importance with Josephson's 1962 discovery of a way in which superconducting phase differences could be directly observed in the laboratory. The class of devices in which this can be done are called Josephson junctions. They come in a variety of forms, but they all achieve a weak coupling between two superconductors. The supercurrent which can flow between the two is a periodic function of their relative phase, so that this relative phase becomes directly observable.

Superconductivity has been shown to occur in several thousand different metallic elements and alloys, and even a few semiconductors. In the beginning, the search for new superconductors was simply a matter of checking readily available materials. Now, however, the searchers are armed with the insight gained from the BCS theory and with empirical rules relating the appearance of superconductivity to such parameters as the number of electrons per atom and the crystal structure. As a result, new materials are being made and studied, and the total number of known superconductors is increasing at a rapid rate. The range of desirable characteristics like transition temperature and critical field, while increasing, is not increasing so fast. One of the most important unanswered questions remaining in superconductivity is whether it is possible for the phenomenon of superconductivity to exist with significantly higher transition temperatures than are now known. A considerable increase in transition temperature would obviously have far reaching technological implications. For the moment the highest known transition temperature is about 21 K, with a corresponding upper critical magnetic field in excess of 400 kG. The totality of known superconducting materials provides a sufficient variety of superconducting, mechanical, and metallurgical properties for a host of technological applications.

Let me now briefly describe some of the more notable devices which have been developed, using the effects and materials I have just described. First of all, there is a large class of devices which may be said to comprise the area of superconducting electronics. The earliest devices in this category were based on the "classical" properties of superconductors, zero resistance and the Meissner effect. More recently, a host of devices have been developed in which the phase coherence of the superconducting state plays a dominant role. These invariably incorporate some type of Josephson junction.

Among the "classical" devices are galvanometers and voltmeters. These are especially attractive for detecting small dc voltages from sources at cryogenic temperatures where the intrinsic source noise is very small and is likely to be overwhelmed by detector noise in any room temperature detector. The most sensitive devices of this sort use mechanical or thermal modulation of the resistance or inductance of a superconducting circuit to convert a dc voltage to an ac signal. Voltage resolutions of the order of 10^{-14} V with time constants of about 1 s and source resistances near 10^{-7} Ω have been achieved.

The magnetic field produced by a current flowing in one circuit can switch an adjacent superconducting circuit from the superconducting to the normal state. This is the basis of the cryotron. There was considerable initial interest in the cryotron as a computer element, but it turns out to be too slow for use in modern computers. Cryotron amplifiers have been built with current gains greater than 10^4 and bandwidths greater than 1 MHz. Theoretical estimates of minimum effective noise temperature on the order of 10^{-1} K have been made for cryotron amplifiers.

The extreme temperature sensitivity of the resistance of a superconductor in the transition temperature region has been made the basis of superconducting bolometers for the broad band detection of electromagnetic energy. Noise equivalent powers on the order of 10^{-12} W/$\sqrt{Hz}$ have been achieved. The sensitivity has ordinarily been limited by noise in the room temperature amplifier used with the bolometers. The use of a cryogenic amplifier in conjunction with a superconducting bolometer should give a substantial increase in sensitivity.

The quasiparticle tunnel diode discovered by Giaever consists of two superconductors separated by an insulating barrier less than about 100 Å thick. The appearance of the energy gap in the quasiparticle excitation spectrum of a superconductor causes the current-voltage characteristics for tunneling of quasiparticles to be highly nonlinear. This nonlinearity has been used in quasiparticle tunnel junction oscillators, amplifiers, detectors, and mixers. Tunnel junctions have also been used as alpha-particle detectors in a manner similar to the commonly used semiconductor junction particle detectors. In this application increased energy resolution is obtained because of the small energy gaps in superconductors compared with semiconductors. Quasiparticle tunnel junctions have also been used as generators and detectors of very high frequency phonons.

The low loss characteristics of superconductors makes possible rf resonant circuits of very high Q. Q's up to 10^{11} at 10 GHz and 2×10^8 at 30 MHz have been reported. These Q's are on the order of 10^6 times larger than those achievable with normal conductors. Low loss miniature delay lines have also been developed.

Devices based on Josephson junctions (quantum phase devices) appear to be generally more promising than corresponding classical devices. One of the most important quantum phase devices is the quantum interferometer. This consists of a superconducting loop containing one or more Josephson junctions. The periodic dependence of the junction current on the phase difference across the junction(s) together with phase coherence around the loop results in a periodic variation of the circulating loop supercurrent with magnetic flux linking the loop. The period is the flux quantum. The state of the circulating supercurrent can be probed by dc or rf methods. The device is an extremely sensitive magnetometer or, if coupled to a coil (with series resistor), a galvanometer (voltmeter). The flux sensitivity which has been achieved is on the order of 10^{-4} - 10^{-3} flux quanta, and this translates into field sensitivities of 10^{-10} G or better and voltage sensitivities on the order of 10^{-16} V. Theoretical considerations suggest that some improvement on these sensitivities is possible.

If biased with a dc voltage V, a Josepshon junction carries an oscillating supercurrent with fundamental frequency $\nu = 2eV/h$. This is the ac Josephson effect. It means that the junction is an active as well as a nonlinear rf device. It can be used as a detector of electromagnetic radiation in several modes of operation, and should be useful at frequencies ranging up through the far infrared. Achieved noise equivalent powers are 10^{-14} W/$\sqrt{Hz}$ at 300 GHz and 5×10^{-15} W/$\sqrt{Hz}$ at 100 GHz.

Because of the ac Josephson effect, a Josephson junction can be viewed as a frequency-to-voltage transducer with transduction ratio $2e/h$. The effect has been used to determine $2e/h$ with an accuracy several orders of magnitude greater than is possible with other methods. It has also been shown that the effect can be used to provide a voltage standard with a stability and precision (parts in 10^8) at least an order of magnitude better than those available from the electrochemical standard cells which are currently used as basic voltage standards throughout the world.

If the voltage bias is derived from a source with a small

resistance, thermal noise in the resistor will frequency modulate the Josephson oscillation. Measurements of the resulting frequency spectrum yield a measure of the absolute temperature of the resistor. Hence a Josephson junction can be used as an absolute noise thermometer. Excess-noise-limited precisions of better than 5 mK have been reported in the millikelvin temperature range.

The current-voltage characteristic of a Josephson tunnel junction has, for currents less than the Josephson critical supercurrent, two possible states for a given current, one at zero voltage, and one at a voltage ($\sim$ 1 mV) determined by the superconducting energy gap. The switching time between the two states has been shown to be less than a nanosecond. The Josephson tunnel junction is thus a bistable element with potential computer applications. Information storage and processing devices based on the transfer of flux quanta between different portions of extended circuits containing Josephson junctions have been proposed. Research aimed at the realization of practical computer elements is under way in several laboratories.

The superconducting electronic devices I have just described are all characterized by small physical and electrical parameters. They belong to what might be called microsuperconductivity. There is a contrasting area which might be called megasuperconductivity, the technology of very large scale superconducting devices. The most advanced area in superconducting technology belongs here, the technology of superconducting magnets. The high current carrying capacity of Type II superconductors in high magnetic fields has led to the development of very-high-field large-volume magnets of many kinds. Laboratory magnets producing more than 100 kG in volumes of a few cm^3 are now common. Very large magnets have been built for bubble chambers and bending and focussing magnets in high energy physics laboratories, and for use in plasma physics and controlled fusion research. Some of these magnets are characterized by numbers like 20 kilogauss in a 4 meter diameter, with total energy storage capacity approaching 100 megajoules.

Other potential large scale applications of superconductivity include ac and dc motors and generators, electric power transmission lines, magnets for energy storage, magnetically levitated trains, and particle accelerators. Theoretical analyses of all of these applications have been made, but development of actual prototype devices has been rather limited because of the high cost of experimental work on the large scale necessary for realistic assessment of designs. Small superconducting motors and generators have been built in several laboratories, and a 3250 horsepower (2.4 MW) motor has been successfully tested in the United Kingdom. Several laboratory test facilities exist for determining superconducting transmission line operating parameters. Some preliminary experimental work related to magnetic levitation and propulsion for trains is currently under way. Work on large superconducting particle accelerators is in progress in laboratories both here and abroad.

That is a brief list of what we now have at our disposal in superconducting science and technology. Now let us consider what we can probably do in superconducting technology during the next decade. For the moment, let us consider only the technical possibilities and ignore the limitations imposed by money and manpower considerations. I believe reasonable extrapolations from the present state of the art indicate the following: In the area of superconducting electronics, we can build magnetometers with sensitivities of 10^{-11} G/$\sqrt{Hz}$ or better. This great sensitivity facilitates the construction of very sensitive gradiometers and systems for the detection of higher order magnetic field derivatives. Such devices will become increasingly important in applications of superconducting magnetic field sensors outside carefully shielded laboratory environments, e.g., for the detection of geomagnetic anomalies in the field. The ambient low frequency magnetic noise at relatively isolated locations on the earth's surface is in the 10^{-8} - 10^{-5} G range, and it can be considerably greater in populated areas. The use of derivative sensing devices together with appropriate data processing and filtering techniques will permit the exploitation of superconducting magnetometers to study magnetic fields due to a variety of interesting man made and geophysical sources.

The superconducting magnetometer can be used to measure very small susceptibilities in very small samples in very small magnetic fields. It has already been exploited in this way to measure the fluctuating susceptibility of a superconductor near its transition temperature, and the susceptibilities of biological materials. It is to be expected that such applications will increase in number.

The use of a magnetometer to detect the field produced by a magnetic or superconducting mass provides a gravimeter or accelerometer of very great sensitivity. Current work with a superconducting gravimeter has demonstrated a sensitivity surpassing conventional instruments, and further improvements can be made.

A very important feature of the superconducting magnetometer is that it combines high sensitivity with virtually unlimited frequency response. This permits trading off sensitivity for speed in applications where the latter is important. A gradiometer has already been used to detect the magnetic fields produced by mammalian hearts and brains. This opens up new possibilities in biomedical studies and diagnosis. Communication by means of very low frequency electromagnetic fields (10 to 10^5 Hz) is of great importance when the receiver or transmitter (or both) are immersed in the sea or are far underground. The appreciable conductivity of these media prohibits transmission of radiation in the conventional communication bands. Receivers based on superconducting magnetometers can provide significant improvements in sensitivity and hence communication range and reliability.

The voltmeter version of the superconducting magnetometer can be used in a greater variety of applications than it has seen so far. It is presently sensitive enough to be limited by helium temperature Nyquist noise in resistances below about 10^{-1} Ω. It will be most useful for applications involving detection of voltages produced by low impedance devices at cryogenic temperatures, e.g., as a signal readout device for other cryogenic devices.

The sensitivity and utility of the Josephson junction electromagnetic radiation detector can be expected to increase to the point where it is the optimum detector in the millimeter and submillimeter regions of the spectrum. Noise equivalent powers of the order of $10^{-16} - 10^{-15}$ W/$\sqrt{Hz}$ should be achievable. Josephson junctions can also be used as mixers. This mode has not been extensively explored, but there are indications of useful mixing action down to 10 μm wavelength. This mode should become much more widely exploited.

The application of the ac Josephson effect as a basic voltage standard can provide a fundamental international "atomic" standard. This rather exotic application is in fact already well advanced, and the end of the decade should see it well established as the basic standard of voltage in most technologically competent nations. All-cryogenic systems are being developed, and small portable standards with precision and stability equal to those of the national standards themselves can become a reality in the next several years.

Josephson junctions can assume a major role in computer technology. Flip-flop circuits, counters and shift registers based on Josephson junctions have already been demonstrated and their development is being pursued.

The development of high-Q superconducting resonators can provide unique capabilities in many types of rf systems. Development work in this area is furthest advanced in connection with particle accelerators of various types. The production of intense beams of high energy particles is important both in elementary particle physics and in medicine, where beams of exotic particles like pions have distinct advantages in the localized treatment of cancer. Use of superconducting rf systems should permit construction of small relatively inexpensive accelerators suitable for use in major hospitals. High-Q resonators should also have applications in both high and low power rf systems of all kinds, including complex communication systems where filtering and frequency discrimination are important functions of signal processing systems.

In most of the large-scale applications of superconductivity, large-volume high-intensity magnetic fields are the key factor. Large superconducting magnets are already widely used in high energy physics labs, and this use will grow. Superconducting magnets are essential to the economic use of either magnetohydrodynamic power generation or fusion power. If either of these power sources becomes sufficiently practical to be used on a large scale (and fusion power had better, or we're all in big trouble), superconducting magnets can provide the required fields, and, indeed, are the only practical means of doing so.

Superconducting rotating machinery can provide (at least on paper) weight/volume savings of a factor of ten over conventional iron-core machines. This is a major factor even for fixed machines like those in central generating stations, because efficiency tends to increase with size, and present machines are already limited by mechanical strength and shippability. It appears that superconducting generators for central power station use will become economically advantageous for capacities larger than about 500 MW. Most future generators will necessarily be larger than this. The advantages of small size and weight become even more important in vehicle applications, e.g., marine propulsion and aircraft alternators.

Superconducting power transmission lines appear (again on paper) to be the best answer to the future need for very high capacity (> 1 GW) underground lines for transmitting power from remote generating plants into densely populated urban areas. It is not yet clear whether these lines should be ac or dc. Losses due to fluxoid motion are not negligible at power line frequencies, while zero-loss dc operation introduces formidable problems in ac-dc conversion and voltage transformation at the line input and output. However, materials will undoubtedly be further improved and both types of lines can be expected to be economically feasible.

Another attractive large scale application of superconductivity is in magnetically levitated vehicles, specifically, trains. The need for high speed short and medium haul transportation between major urban centers is obvious and not likely to be met adequately by aircraft in the long run. Above 150 mph, wheels become impractical, so that some noncontacting suspension is required, either air cushion or magnetic levitation. The latter has advantages over the former. Levitation is quieter, requires less power, is likely to produce less pollution, and can derive its propulsion from the same magnet system used for levitation by operating it as a linear induction motor. A practical magnetic levitation system can be developed based on either normal or superconductors. It should be noted that the Japanese National Railway is committed to the operation of a magnetically levitated train by the end of the coming decade. It remains to be seen whether the system will be normal or superconducting.

All of these large scale devices can be built and operated at least in prototype form during the coming decade. The extent to which this will actually occur will depend on the financial support obtainable for large scale experimentation, not on technical factors.

I believe we can expect some further progress in superconducting material technology. Transition temperatures will rise to perhaps 25-30 K, and materials will be developed with electrical and mechanical characteristics suited for the applications I have mentioned. I do not expect to see a liquid nitrogen temperature or room temperature superconductor within the decade, at least not in a form suitable for technological application. It is important to emphasize, however, that the materials we already have in hand are adequate for the technology we are likely to be able to develop during the coming decade.

Now what should we do with superconducting technology? The answers to that question are very much a matter of personal opinion. Here are some of mine. First of all, I believe that superconducting technology will have to put up or shut up during the coming decade. By 1981, we will know whether superconducting technology is going to be a major factor in the world's economy and technology or merely a source of unique specialized laboratory instruments. We will either have prototype or operational superconducting rotating electrical machinery going into power plants and air, land, and sea vehicles, magnetically levitated trains in prototype or operational status, cryogenic computers in being, superconducting magnetometers in use on land or sea, hospitals with magnetocardiographs and superconducting particle accelerators, and communication systems incorporating superconducting sensors and filters, or we will have concluded that superconducting systems do not provide practical solutions to real world problems. I believe superconducting technology does have a major role to play, and I believe we can and should actively pursue all of these avenues.

It would be nice if, having proclaimed our confidence in the ultimate utility of our science, we could simply proceed directly to the development of the technology. Unfortunately, it will not be that easy. A large conventional electric generator, a machine which utilizes a mature technology based on more than half a century of development, costs in the neighborhood of 20 million dollars. It is obvious that a development program aimed at replacing that machine with a superior superconducting one will require funding at a level which is a significant fraction of the total current annual investment in the entire field of superconductivity. The same can be said of the other large scale applications. Even in superconducting electronics, the development costs of actual systems incorporating superconducting devices are likely to be considerably larger than typical support levels for a basic research project in superconductivity. If we are to marshal that kind of support, we must clearly identify unfilled needs which can be satisfied best or perhaps only by superconducting devices. This requires establishing effective communication with the potential ultimate users of our technology so that we can tell them what we can and cannot do, and they can tell us precisely what most needs to be done. This task of mutual education and definition and redefinition of goals must proceed concurrently with the development of the science and technology, and is equally as important. It is essential that scientists in the field address themselves to both problems.

It should be apparent from my brief description of the status of our understanding of superconductivity and from the other chapters in this volume that the science of superconductivity is fairly well developed. We do not as yet have any extensive experience or capability in the technology, with the single exception of superconducting magnets. We are thus long on science and short on technology at this point, and a great deal remains to be done before we have a clear idea of the true dimensions of the technology. There is a great deal of hard work and opportunities for many a slip twixt laboratory and loading dock or field station. Superconducting systems must be designed for operation under difficult industrial and field conditions by personnel not endowed with a Ph.D. in low temperature physics. Prototypes must be built and exhaustively tested. Ways must be found to fabricate in quantity such devices as high-Q cavities and Josephson junctions with long term reliability and performance comparable with that now achieved by long-suffering graduate students in laboratories. We need materials suitable for large scale load bearing structures which can operate at cryogenic temperatures and across the gradient to room temperature with acceptable thermal characteristics. One area which obviously will require a great deal of attention is refrigeration. The periodically refilled dewar suffices in the laboratory but it will not in most applications. We need a variety of refrigeration systems ranging from large fixed closed-cycle systems for cooling large electrical machinery to simple essentially zero-power systems for cooling Josephson junction devices without introducing mechanical vibrations or magnetic fields. This list of needs can be extended almost indefinitely.

Who is going to do all this? At the moment, the knowledge and expertise in superconductivity is largely in the hands of research scientists. Some of these will take an active part in the development of the technology, and some may transform themselves into superconducting engineers. But most lack the experience in solving real engineering problems which will be needed. It is clear that if superconducting technology is to progress expeditiously, new groups of physicists, metallurgists, engineers, etc., must be brought together and must educate each other in the techniques necessary to achieve their goals. This is already taking place in many places, and we can expect to see more of it. Until it develops into a well defined discipline itself, superconducting engineering will be a truly interdisciplinary activity.

How is all this to be organized and paid for? Many of the small scale electronics applications will continue to be followed through their initial stages in university and industrial research laboratories. But what we really must have are complete systems combining superconducting devices, refrigeration, electronics, etc. These are beyond the capabilities and purses of most present superconductivity research groups. In the area of large scale applications, generators, motors, trains and the like, it will be difficult for most laboratories to do much of anything without significant changes in structure and funding. It is clear that private industry and the <u>development</u> funding parts of the federal government will have to assume the responsibility for bringing together the talent and money to define the job and then get it done. There is nothing here that has not already been done many times with other technologies, but I would like to emphasize again the crucial importance of education, communication, and efficient information transfer. I hope this volume will contribute significantly to this process.

In summary, the picture I have tried to paint here is that of a well developed and broadly based science, with a very exciting technology in the wings. I hope and believe that the performance will vindicate the advance notices. If it does, superconducting technology will have an economic impact which will dwarf the investment we have made in superconductivity research and satisfy the most "relevance" minded that it was all worth while. Let us hope that we can, in Einstein's words, "learn to make a sensible use of it."

I would like to thank many of my colleagues, particularly J. Clarke, W. A. Little, J. E. Mercereau, P. L. Richards, and D. J. Scalapino for their contributions to the formation of the opinions I have expressed here. I am indebted to Judy Goodstein for bringing the Einstein address to my attention.

References

1. R. W. Schmitt and W. A. Morrison, in *Superconductivity in Science and Technology*, M. H. Cohen, Ed., University of Chicago Press, Chicago, 1968. I also commend to the reader's attention the other articles in this volume.

2. Superconducting Devices and Materials, a quarterly literature survey compiled by W. S. Goree, E. H. Takken, R. A. Kamper, and N. A. Olien and published jointly by the National Bureau of Standards and the Office of Naval Research.

3. I have not attempted to compile a comprehensive set of references to the list of phenomena and devices which follows. Appropriate references may be found in the other chapters in this volume.

A SUMMARY OF THE COURSE

J. Bostock

Institut für Theoretische Festkörperphysik
Universität des Saarlandes
Saarbrücken, West Germany

After Dr. Langenberg's masterful talk on superconducting technology and its implications for the coming decade, my job of giving the summary talk is immeasurably easier. Dr. Langenberg has reviewed most of the technological aspects of the subject and given us some insight into the probable future of their development. Since we have had the experts in the various areas of superconductivity come and talk about their work, I will not attempt to review each and every contribution to the meetings. What I hope to do is to briefly outline these sessions, indicate where various applications fit into the general area of superconductivity research, and point out some of the areas of this research which particularly interest me - in terms of problems to be solved, big and small, either by the experimentalist or by the theorist.

My task, to give a summary of this summer school on superconductivity, is not an easy one for several reasons. First, the course has lasted 12 days and we have seen that progress is going forward on a very broad front in a wide variety of phenomena associated with both basic research and applications. Of necessity then, I can only comment selectively on a few subjects which mirror my own personal interests.

However, I should emphasize that meetings such as these which bring together recognized experts on both the scientific and the technical aspects of superconductivity are truly valuable. Previously, such meetings have been almost exclusively devoted to fundamentals and basic phenomena, yielding a climate rather remote from the practical aspects. But in recent years a new and powerful force has been demanding contributions from basic scientific research and technology. I refer, of course, to the increasing emphasis on the social problems of mankind: pollution, health, transportation, communication, power sources, etc. Here in this course an attempt has been made to bring into better focus the interaction between the "scientific'and "real" worlds, not only to provide answers to some of our more pressing social problems, but also to provide a cross-fertilization between the research and technological aspects of superconductivity.

It would also seem from various discussions that have been held during these two weeks that the immediate future of "superconductivity engineering" still lies with the physicist. There are two reasons for this. One, it is the physicist who has the most pressing need for its services; the most spectacular examples of superconductivity at work are in high energy physics and plasma research. And two, in most of the present applications superconductivity is competing with a long established technology that has been developed to a very high standard of general efficiency and reliability and consequently low manufacturing costs. This becomes particularly evident in the discussion of applications concerned with computer technology, levitation devices for transportation, and motors and generators. Not only is it necessary to develop the various superconducting components for these applications, but the low-temperature support technology is still largely unavailable, although from a cost standpoint this is not the prohibitive feature of these systems. Perhaps the largest single application of superconductive technology, which also appears to be the most promising one, is to power transmission lines, for reasons of both economics and need.* (1)

The second reason why giving the summary is difficult is implicitly contained in my short digression on the applications of superconductivity. It is clearly impossible to report on the various "extra sessions" and informal discussions. The subjects of these meetings, ranging from the economics of applications to advanced laboratory techniques, form a perfect supplement to the more formal discussions; and I suspect nearly half the business of the course was conducted during these get-togethers. Certainly my own opinions on the economics of applications, for example, were formed and informed there. Also, I'd like to add that my awareness of some of the problems that I shall suggest is directly traceable to these conversations and the generosity of the people who originally suggested them to me.

That brings me to my last problem. I have to emphasize that I am certainly not an expert in all the areas of superconductivity we have covered, neither in technology nor in basic research. I have worked on a few small problems in the general area of superconductivity and I know that sometimes only the very difficult work remains to be done. Nevertheless, I think we have now passed

A SUMMARY OF THE COURSE

beyond the use of very simple (and successful!) models of superconductivity, so that, whereas most of the theory we have studied has been concerned only with the electron-phonon interaction, I shall have to suggest problems which seem much more sophisticated when viewed against this background. Similarly for the experimentalist, I may suggest problems that require, perhaps, a great deal more expertise than those we have discussed here. But then, this course is testimony to how developed the field has become.

As a prelude to a detailed discussion of certain topics, I'd like to briefly outline the route taken during the course. The directors introduced the entire subject of superconductivity, discussing the history of the field, the salient characteristics of superconductors, the classical calculations which give a good description of some of the behavior of superconductors, and the physical bases of both the phenomenological and microscopic theories of superconductivity. This introduction was followed by a formal, mathematical development of the microscopic theory. This approach relies heavily on standard Green's function methods. Dr. Mathews's lectures on this method provide a general and rigorous account of such procedures.

Perhaps, considering the unfamiliarity of this approach for most of us, we should review the arguments which show why this method is essential for the proper phrasing of the interesting questions about superconductivity. The first and most basic question about any physical phenomenon concerns the essence of the effect. The BCS theory was eminently successful in pointing this out for superconductivity. A second question, of great interest to theorists, is that of the best language to use for the description of the effect. Here one would like to construct a general formalism that allows one to study the response of a typical system to relatively simple perturbations (changing the number of particles in the system, for example), and in so doing predict the macroscopic behavior of the system. The usual procedure for these calculations is to apply statistical methods to the problem. For many-body problems the formal similarity between the quantum-mechanical time-evolution operator, $\exp(-iHt)$, and the statistical operator, $\exp(-\beta H)$, is exploited in the formalism of thermodynamic Green's functions. In the case of superconductivity this technique is particularly valuable for two reasons. First, it avoids the Landau quasiparticle approximation inherent in the BCS solution to the problem; i.e., we are no longer confined to considering a set of independent stable quasiparticles. With the Green's function method we can treat spontaneously decaying excitations which interact with one another. Second, the net interaction causing superconductivity is not describable by an ordinary two-body potential since it is a co-operative interaction in which the polarization of the medium as a whole plays an essential role. The Green's function method allows us to include any general time-retarded interactions, involving the medium as well as the single excitation, in the description of the many-body state. Of course, the success of the method depends critically on using the appropriate interaction - and that is quite another problem!

The nature of the superconducting state and the formalism used to describe it having been presented, various aspects of experimental superconductivity were then discussed by Drs. Gregory, Schooley, and Deaver. Dr. Schooley reviewed several approaches to determining transition temperatures and methods for enhancing them for superconductors. Some of the methods used in thermometry were, by the way, quite shocking to some of us. I refer especially to the very economical "Noise Thermometer" using SIN junctions. (2) Dr. Gregory concentrated his efforts on three effects: boundary scattering, (3) the ac-susceptibility techniques, and radiation from tunneling structures. Dr. Deaver discussed probably the most fascinating experimental aspect of superconductivity, the Josephson Effect (or to be more accurate, Effects). This was a particularly enlightening exposition of the subject, based on simple physical models and familiar electrical analogs. Taken together with the discussion of millimeter and submillimeter detectors and devices given by Dr. Shapiro, the full spectrum of the experimental manifestations of the Josephson Effects were presented.

I would like to digress here for a moment to point out the unique opportunity afforded at this course to go into a working laboratory and see not only the "How?" and "What?" of various experiments, but also the detailed considerations involved in designing these experiments and interpreting the results obtained from them. Let me choose just one of the experiments that Dr. Gregory discussed during the advanced laboratory sessions, tunneling phenomena. Surely most people would agree that this subject is in such an elegant state that we can concentrate on other fields of research. However, we have seen here the difference between tunneling curves of ideal single-gap materials and those of multiple-gap superconductors displaying signs of local gap anisotropy (Figure 1). (4) As Dr. Gregory also pointed out, some very detailed theoretical work is still necessary to correlate the experimental results with band theory of real materials. (4) In fact, further research, both theoretical and experimental, is still required in this area, not only to refine our knowledge but also to explain new observations. (4, 5)

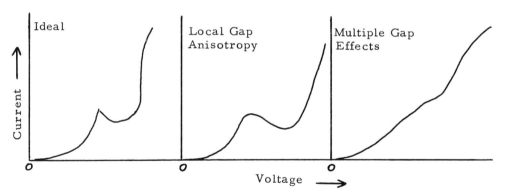

Figure 1. Characteristic Tunneling Curves

The application of theory to various specific areas was undertaken by Drs. Bloch, Koonce, and Scalapino. Dr. Bloch's derivation of the Josephson Effect in a superconducting ring (6) is a closely reasoned, persuasive discussion. The arguments are drawn from basic physics, quantum mechanics, and density matrix theory. (7) Due to time limitations, Dr. Bloch was only able to sketch further developments in the theory, such as interposing a barrier in the ring. I urge you to read the rest of this theory in his recently published work. (6,7) It is well worth your effort, and the clarity of presentation is equal to that of the talk we heard this past week. The theory of enhancement effects in superconductors and the theory of semiconducting superconductors presented by Dr. Koonce gave us an illustration of the way in which the formal theory of superconductivity, introduced by Dr. Mathews, is used to obtain concrete results. And finally, Dr. Scalapino discussed the phenomenological approach to time-dependent superconductivity. His description of a non-equilibrium condition in the "Josephson device" is to be contrasted to the equilibrium situation described by Dr. Bloch.

The characterization and classification of superconducting materials was taken up by Dr. Rose (type II's), Dr. Matthias (elements and compounds), and Dr. Hein (A-15 compounds). This review of material properties, together with the first chapters on theory and experiment, constitute the background necessary for an appreciation of the remaining sessions, those devoted to the technological aspects of superconductivity. Dr. Langenberg's article summarized in some detail all those applications which we have discussed, so I shall proceed to an interpolation and extrapolation of some of the areas I have just mentioned.

Although the best place to start in any discussion of open-ended problems in superconductivity might seem to be "mechanisms which lead to superconductivity", this would not be very fruitful since, with one exception, I can only tell you about those which do not lead to superconductivity. (Perhaps one should add a word of encouragement at this juncture for those brave souls who <u>do</u> attempt first principles calculations with new mechanisms. The results they get always seem quite a long way from experimental verification, but eventually, if we wish to understand what superconductivity is all about, calculations of this kind have to be done. Only then can predictions be made, which is, of course, the ultimate goal of all theory.) In fact most of the speculative theoretical papers deal with additions to or enhancement of the naturally occurring phonon-induced electron-electron interaction. (9) Dr. Koonce discussed this subject quite thoroughly, so I'll refer you to his article. However, I would like to make a few additional comments on his discussion of the McMillan equation (10) for the critical temperature of a superconductor.

The McMillan equation for the critical temperature, derived using the Green's function formalism, is:

$$T_c = \frac{\theta_D}{1.45} \exp\left[\frac{-(1+\lambda)}{\lambda - \mu - 0.62\lambda\mu}\right].$$

This can be compared to the BCS formula,

$$T_c = 1.134 \; \theta_D \exp(-1/N_o V),$$

where $N_o V$ is just the "effective attractive interaction" in the superconductor, $(\lambda - \mu)$. First, in considering the lattice polarization by electrons we neglected the inherent <u>Coulomb repulsion</u> between electrons. This repulsion will cause <u>a reduction of the "net BCS interaction"</u> by a factor of $(0.62 \lambda \mu)$. Secondly, we know that quasiparticles in the many-body system are <u>electrons interacting with the lattice</u>, not bare electrons, and thus have a higher effective mass than bare electrons $[m^* = (1+\lambda)m]$. This means that the effective interaction in the system has been overestimated and must be reduced (or renormalized) by a factor of $(1+\lambda)$. The advantage of having such an equation is that given the normal state properties of a material we should be able to predict whether or not it is superconducting via the electron-phonon mechanism. As was pointed out earlier, the problem is that we do not know the appropriate normal state parameters, nor have we been able to calculate them. Even though this is not a problem in superconductivity, per se, obtaining normal state properties of metals and alloys is an important job still to be done.

We can, however, try to determine T_c as a function of the more common parameters of metals, such as valence, lattice parameter, band mass, etc., using approximation techniques which have been shown to be successful in the past. In particular, I refer you to a series of papers by Seiden where such an analysis of "free-electron-like" materials is considered. (11, 12) To do this theoretically we have to assume "reasonable" forms for the phonon spectrum and zone structure of the system, evaluate the appropriate Coulomb pseudopotential and electron-phonon coupling constant, and finally determine the variations in T_c as a function of these parameters. For the experimentalist some compound or alloy system should be chosen where one parameter can be varied and the results compared to theory. It may sound complicated, so I hope you will read these papers for insight into the physical bases of the variations in T_c in simple superconductors.

We have also heard some discussion during the course concerning predictions of the McMillan theory for T_c of "softer-systems" of known superconducting materials. This refers to an analysis made by McMillan of T_c as a function of the average phonon frequency of the material, using

$$T_c \sim \theta_D \exp(-[1+\lambda]/\lambda),$$

where the electron-phonon coupling constant can be written

$$\lambda = N_0 <\mathcal{J}^2> / M <\omega^2>.$$

Here M is an average ionic mass, N_0 is the density of states at the Fermi surface, $<\omega^2>$ is the average square phonon energy and $<\mathcal{J}^2>$ is the average over the Fermi surface of the squared electron-ion matrix element. As Dr. Koonce discussed, McMillan found that $N_0 <\mathcal{J}^2>$ was remarkably constant for some of the different classes of materials he investigated. Using this dependence of λ on the phonon frequency, McMillan then showed that there should be a maximum in T_c as a function of phonon frequency for given classes of materials. (Figure 2, taken from Ref. 12.) This is because a very soft lattice enhances the electron-phonon interaction - the response of the ions to the electrons is very large. Then as $<\omega^2>$ increases, λ decreases, and T_c should increase. On the other hand, as McMillan also pointed out, if the lattice is softened the Debye energy becomes smaller, and since electrons scatter only to states within a Debye energy of the Fermi surface the number of states available for scattering decreases. Thus, even though the interaction becomes stronger, the available phase space decreases, and T_c goes through a maximum. Dr. Koonce has discussed the attempts to maximize T_c in a given material group, both the successes and failures.

I would like to bring to your attention some experimental results which deviate from the constant $N_0 <\mathcal{J}^2>$ rule. (13)

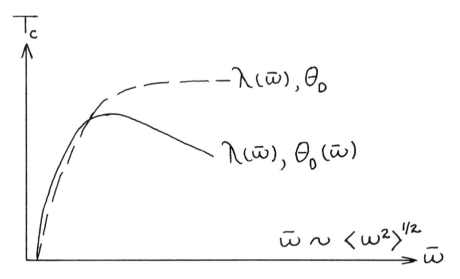

Figure 2. The dependence of T_c on average phonon frequency.

One particularly striking example was observed by Dynes in a tunneling study of the Indium-Thallium alloy system. (13) (See Fig. 3) Here we see the measured T_c values and below them the calculated values of the product $N_0 <\mathcal{J}^2>$. Although the variation in $N_0 <\mathcal{J}^2>$ is not as great as the variation in T_c, the correlation is unmistakable. Does this phenomenon have a wider applicability to other alloy systems and series of compounds? For those of you who have tunneling data on such systems, I would suggest an analysis of your results to see if such a relationship between T_c and $N_0 <\mathcal{J}^2>$ exists. To the theorist I would ask these questions: (1) Can we explain why this is happening?
(2) Can we develop a formalism for calculating this effect? As was mentioned previously by Dr. Koonce, a start on this second problem has already been made by trying to calculate the electron-ion matrix element for transition metal superconductors. (14) On the one hand, it has been suggested that in these systems superconductivity is due to a phonon-induced d-p electron coupling via an incoherent scattering of electrons off statically displaced ions. (14) On the other hand, it has been argued that it is the interband d-d electron scattering off coherent lattice displacements that causes the superconductivity. (14) Clearly more work needs to be done on calculating the electron-phonon coupling constant from first principles.

To improve our model of phonon-induced superconductivity, we will have to take into account the details of the structure of real

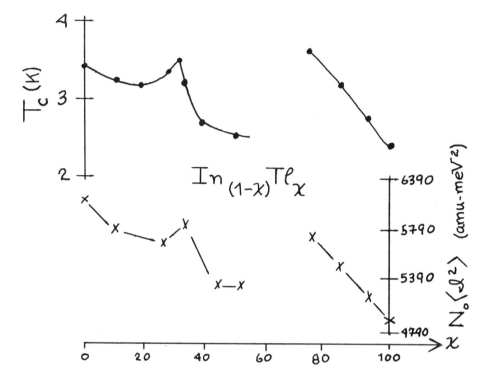

Figure 3. Comparison of T_c (•——•) and $N_o \langle \mathcal{J}^2 \rangle$ (x——x) vs Tl concentration for the In-Tl alloy system.

materials. (15) Let me mention only one example of the applicability of such work. Havinger has noted an oscillatory behavior of T_c as a function of the number of electrons per atom in several alloy systems. (16) He suggests that this behavior can be explained in terms of Brillouin zone effects, at least for cubic structures. That is, there appears to be an enhancement or reduction in T_c whenever a Bragg plane intersects the Fermi surface. This is an interesting idea that should be explored further.

I don't want to leave you with the impression that we know everything there is to know about the electron-phonon interaction when we assume an ideal isotropic one band model of superconductivity. Two recent papers point this up rather well. Cooper and Stölan discuss the origin of the pairing interaction by using the Goldstone diagram technique to explain why the $-\vec{k}\downarrow$ and $\vec{k}\uparrow$ pairing dominates the interactions in the "BCS shell" surrounding the Fermi surface. (17) Nam has shown that contrary to the predictions of the BCS model, there does exist a finite spin paramagnetic susceptibility of electrons in superconductors at absolute zero, and that even in the weak-coupling limit, it does not vanish. (18) Bardeen has conjectured that this effect could be due to the van Vleck magnetism, (18) but as Nam points out, this is still to be proved. Another fundamental question would be, how good a parameter is the pseudopotential, μ , for describing either the superconducting or normal state of a material? A systematic application of the McMillan formula for T_c shows that the results are not particularly sensitive to this Coulomb term (except, perhaps, at very small coupling constants) and in fact μ , itself, does not vary much for very different types of materials. (10) Perhaps μ is too coarse a parameter for describing metallic states. But, if this is true, what is the proper parameter?

It has become apparent that one of the most important areas in experimental superconductivity is that of materials. Dr. Rose gave a complete description of the metallurgy of type II superconductors, especially those which are already considered "practical materials", and of some that he considered potentially practical ones. His chapter is a well-organized and well documented review of the subject. The applicability of this information was continually reinforced in every succeeding talk involving an application of superconductive technology to other fields, such as plasma research, motors, and magnets. The picture which emerged was that the usefulness of a superconducting material depends crucially on how it can be fabricated. It also became evident that only by understanding the diverse phenomena already seen will we be able to make materials in the future that have the characteristics necessary for producing given effects. And in fact, the lack of favorable materials is the major roadblock to progress in many technological areas. I gather that a ductile superconducting alloy with a reasonable critical temperature, a very high current carrying capacity, and an upper critical field in the 100 - 150 kgauss range would be a start in the right direction.

Experimentally, the dynamics of flux flow in type II materials needs to be investigated, as was so clearly pointed out. On the other hand, Dr. Rose reminded the theorists that even now we do not have an adequate explanation of pinning forces and mechanisms in the type II materials. It is true that the magnetic properties of these superconductors have been studied extensively by many researchers, and at the present time a relatively good "intuition" exists concerning the sources of hysteresis and magnetic instabilities. (19) However, new effects continue to be observed (20) and neither theoretical nor empirical models have been found which can fully describe or explain more than limited amounts of published data. (20, 21)

Of course the other area of investigation into materials concerns enhancing T_c in known superconductors and in searching for new superconductors. Dr. Matthias and Dr. Schooley both remarked that here, too, one takes an empirical and somewhat metallurgical approach. The search for high T_c superconductors has led to a listing of all material properties and a search for some pattern of behavior in terms of, say, electron/atom ratio, atomic volumes, melting points, and so on. (Dr. Hein has given us quite a complete list, you remember.) Each experimentalist has his own favorite, of course! But as previously noted, a superconductor having only a high transition temperature, without other properties such as ductility, high H_{c1}, high current carrying capacity, etc., would not produce much effect in superconductive technology.

A major result from Dr. Matthias's work has been to show that so far the best way of making high T_c materials is by making them softer, more unstable, and more easily polarizable. However, most substances that one would expect to be high temperature superconductors have inherent crystalline instabilities; that is, the crystals are metastable at best. Not only that, but these instabilities increase until eventually the crystal won't even form in the first place. (1d) Dr. Koonce also pointed out the main reason for this: the electron-phonon interaction which leads to pairing of the electrons at the Fermi surface also lowers the phonon energy, and the condition that phonon energies be positive is a condition for the stability of the lattice (i.e., if you make the electron-phonon interaction very large, then you are also making the phonon-electron-phonon interaction very large until finally it distorts the lattice to such an extent that a phase transition takes place). Hence, for phonon-mediated superconductivity the more general limit to T_c is given by the stability of the crystal, itself. This particular criterion, however, has never been made quantitative! It would seem that for T_c's between 22 and 25 K these observed "metastabilities" are still sufficiently long-lived to cope with; (12) and A-15 compounds are good examples of this type of behavior. Matthias's advice to the young experimentalist would be to search for high T_c's among materials having metallic phases that should never form in the first place, so that they are relatively unstable, and which arrange themselves in a cubic environment. It might also be wise to try to get an electron concentration in the range of 4.5 to 4.8 electrons/atom.

To the theorist Dr. Matthias would suggest that a study of the inter-relationships among ferroelectricity, ferromagnetism, and superconductivity might afford some insight into possible mechanisms for superconductivity or for its non-appearance. (22) And, as he is so fond of saying, perhaps we should exchange the study of one complicated phenomena, superconductivity, for another, that of melting points - since the anomalies of the latter seem to be intimately connected to the former.

The discussion of high T_c materials always brings into play the A-15 structure compounds. Dr. Hein presented a lucid and stimulating review of properties of these materials, which taken as a whole have the highest known T_c's. As he indicated, there is still quite a lot to learn about these materials, especially concerning their T_c's. (23) (By the way, you may hear a little bit more than you want to on this subject - but after all, I am working in the field.) What experimental data is lacking, is often due to the difficulty in fabricating these materials because of their brittleness and to the instability of many of these compounds in a stoichiometric composition. The phase diagrams of many of them are not even confirmed.

As Dr. Hein mentioned, the model which has been most successful in describing the properties of the A-15 compounds is a one-dimensional tight-binding model. (24) Here the assumption that the Fermi level lies close to a d-electron band edge which has a very steep density of states gives rise to the high T_c, but also yields the instability which causes some of these crystals to undergo a phase transition from the cubic to the tetragonal structure. And it is just those compounds in the series which undergo the transformation that have the highest critical temperatures. It is a direct electron-acoustic phonon coupling (and thus an instability involving only acoustic phonon modes) in this model which causes the transformation. Although there have been a number of experimental and theoretical studies published on this phase transition, nevertheless fundamental questions concerning the nature of the transformation remain. When these so-called martensitic transformations were first observed, Anderson and Blount suggested several possibilities for mechanisms of the transformation. (25) Prominent among these was that the transformation might be driven by an optical-phonon instability. It has recently been shown by Shirane and Axe (neutron scattering) that indeed an optical-phonon mode is involved in the transition and that this mode is linearly coupled with the soft shear acoustic mode previously observed in these crystals. (26) However, their data does not require an actual optical instability. So here we have another theoretical problem to be solved.

It is also interesting that for the one-dimensional model it has been shown that there should be a strong transverse acoustic phonon coupling to the d-electrons, as well as the garden-variety longitudinal coupling. (27) This, of course, will also give rise to an enhanced T_c for these substances. This calculation has implicit in it that the coupling constant for the A-15's would also be temperature dependent. (27, 28) This shows just how different these substances are and why they have received so much attention in the recent past.

There is also much experimental work that needs to be done on these compounds. We do not have very much evidence on just exactly what the energy gaps are in these materials, for example. (29) Getting this information is not so easy as you may think, however, since they form poor surfaces and have a coherence length of the order of only 20 Å It would be interesting to see the electronic contribution to the specific heat in the normal state at very low temperatures (where the A-15's are ordinarily superconducting) and to compare transforming and non-transforming crystals. And lastly, the energy-dispersion curves for low temperatures and low wave numbers should be determined.

There are many other theoretical questions that have to be answered for the A-15's. First of all, why has the one-dimensional chain model been so successful when clearly there must be some interaction between the chains. Here I should add that, even considering these inter-chain interactions, one does still obtain the peak in the density of states assumed by Labbé and Friedel. (30) And why is it that this one-dimensional model fails to describe the compound Nb_3Al ? (31) In addition, it has been shown that the e/a ratio for V_3Si would be predicted as 4.5 (NOT its actual value of 4.75) and that for Nb_3Al as 4.75 (NOT 4.5). (32) Can this behavior be attributed to an effective change in the potential of the transition metal atom in these compounds, and if so, why?

To get back to the subject of mechanisms for a moment, I would like to mention some recent work of Fröhlich and Rothwarf. (33) Fröhlich postulates that for certain materials (namely the A-15's and the transition elements) there is, in addition to the usual phonon-induced superconductivity, a contribution to superconductivity due to screened plasma oscillations of the "inner (incomplete) bands" of the material. These acoustic plasmons would then be responsible for all the "anomalous" properties observed in these materials. This is an interesting theory and can explain, for example, the data accumulated on Nb_3Sn. However, none of the data establishes unambiguously the existence of the acoustic plasmon mode. On the other hand, is it possible that the high values of T_c found in the A-15's can be attributed directly to the large number of intersections of Bragg planes with the Fermi surfaces in these compounds? (16, 34)

Up until now I have spoken only about the equilibrium conditions in superconductivity. However, we are also interested in non-equilibrium situations, such as the initial response of the material to electromagnetic fields or the effect of time varying fields on superconductors. To do this we need to be able to calculate the point-by-point changes in the Cooper pair condensate. When situations are encountered in superconductivity where such a function - the order parameter - varies in space, the phenomenological theory of Ginzburg and Landau (GL) is one of the most useful techniques available. Dr. Scalapino reviewed this theory for us and showed how the equations of this theory relate the spatial variations of the order parameter to the vector potential and the charge and current densities of applied fields. However, since the GL theory deals only with thermodynamic equilibrium where each quantity is constant in time, it cannot be applied, as is, to non-equilibrium conditions in the superconductor. On the other hand, the equations of this theory have been shown to follow from the more general Green's function theory. (35) The various extensions of the GL theory, which allow the effective order parameter to vary in time as well as space, are based on either the Green's function formalism, itself, or on the phenomenological GL equations using the time dependence implicit in the more general formalism.

It is interesting to observe in this connection that if we had not modified the Heisenberg representation (by choosing the reference Hamiltonian to be $H - \mu N$ rather than H) we would have found the "anomalous" Green's functions (F and $F^{\dagger}$) to have a time dependence of $\exp(-2i\mu t)$ for equal time arguments. Thus the energy gap function would also have this oscillatory time behavior. (35, 36) For bulk superconductors under time independent conditions this factor is merely an inconvenience, so we transform it away; however, one must restore this time-dependence to the initial equations in any extension to time-dependent calculations. It should also be noted that this time factor contributes to the interference effects between weakly-coupled superconductors. (36, 37)

Whereas a completely general time dependent theory of superconductivity has not been obtained, some progress has been made in extending the GL theory to certain special non-equilibrium situations. (38) From these theories it has been possible to extract some simple phenomenological equations which have been used successfully to treat effects in superconductors occurring near T_c, both in the clean and dirty limits, for type I and type II materials. (39) Dr. Scalapino has presented one such theory here at this course. It is a simple and beautiful model which, by requiring charge conservation in the system, predicts a difference in electrochemical potentials associated with pairs and with quasiparticles. Using this theory, Dr. Scalapino discussed the frequency-voltage relationship for a Josephson tunnel junction, the dynamics of current flow across a normal-superconductor interface, and the dynamic behavior of a superconducting weak link. In this one talk he provided the experimentalist with at least three ways to test his ideas. (Not that any one of the experiments is that easy to do!) Perhaps the most exciting one would be in measuring the deviation of the Josephson frequency relation ($h\nu/2 - eV \neq 0$) in very small volume tunnel junctions. I would also like to emphasize again how very important it is to obtain a more general theory of time-dependent phenomena.

There are many other areas of superconductivity which still need investigation, both experimental and theoretical. A few quick examples would be the mixed state of superconductors, (40) the Kondo effect and superconductivity, (41) and gapless superconductivity. (42) I guess that in general a highly sophisticated and accurate theory of the current carrying states and practical consequences of dissipation in the superconductor has not been derived, either. There is an enormous literature on related topics (all of which are, themselves, very important phenomena) that I have not even mentioned. Despite this I hope that in this summary a feeling for the "superconducting way" as it is practiced in the various branches of the field has been given, and that perhaps I have been able to reflect some of the flavor of these meetings.

So, we have come to the end of this course. We reviewed the present state of the theory of superconductivity and had glimpses of the diverse ways in which the practical art of superconductivity is being applied - very successfully applied, I should say - to the problems of the everyday world: communications, transportation, power transmission, and many other fields. Besides this impressive amount of information, the experts in these various areas have provided us with new problems to be investigated and solved in the future. Perhaps the most esoteric application involves the use of superconducting shielding and a SQUID magnetometer to obtain magnetocardiograms and possibly magnetoencephalograms. (43) It is possible that in this medical use of superconductivity we might find an application for which superconducting technology is essential. It is too early to speculate on this, but it is clear that should this be so, the public demand for the advancement of this technology would increase and facilitate the development of the entire field of superconductivity.

It simply remains for me, not as a reviewer but as a participant in the course, to express our collective appreciation to the people who made these meetings possible. First to the directors who conceived, organized, and ran the course: Edgar Edelsack, William Gregory, and Wesley Mathews. The University staff are to be thanked for their co-operation and unstinting efforts to make the course a success, especially the Physics Department, the School for Summer and Continuing Education, and the Audio-Visual Department. A special vote of thanks should be given to the

Naval Research Laboratory, the National Bureau of Standards, and Dr. Ralph Glover at the University of Maryland for allowing the course participants, en masse, to inspect their low temperature laboratories. I have saved one of our most important benefactors for last, Dr. Alice P. Withrow of the National Science Foundation, without whom so many of the course participants could not have attended. I hope you will all join me now in showing these people our appreciation.

REFERENCES

*References given here are only supplementary to those found in the individual talks. Those talks should be consulted for a more detailed bibliography.

1. General References:

    a. Superconductivity in Science and Technology (Morrel H. Cohen, ed.), University of Chicago Press, Chicago, 1968.

    b. A Guide to Superconductivity (David Fishlock, ed.), MacDonald, London, 1969.

    c. Proceedings of the Workshop on Naval Applications of Superconductivity (J. E. Cox and E. A. Edelsack, eds.), (NRL Report 7302) Naval Research Laboratory, Washington, D. C., 1971.

    d. Journal of Applied Physics 42, No. 1 (1971) and Physics Today 24, No. 8, (1971).

    e. Proceedings of the Fourth Symposium 1969 Spring Superconducting Symposia, (NRL Report 7023) Naval Research Laboratory, Washington, D.C., 1969. Herein books of this series are referred to as "Spring Symposia".

2. R. A. Kamper and J. E. Zimmerman, J. Appl. Phys. 42, 132 (1971).

3. William D. Gregory, Phys. Rev. Letters 20, 53 (1968); W. D. Gregory, et. al., Phys. Rev. B3, 85 (1971).

4. L. S. Straus, et. al., to be published; W. D. Gregory, et. al., to be published.

5. D. C. Tsui, Phys. Rev. Letters 27, 574 (1971); C. S. Koonce, Phys. Rev. 182, 568 (1969); C. Caroli, et. al., J. Phys. C 4, 916 (1971).

6. F. Bloch, Phys. Rev. Letters 21, 1241 (1968) and Phys. Rev. B2, 109 (1970).

7. F. Bloch, Phys. Rev. 166, 415 (1968) and Phys. Rev. 137, A787 (1965).

8. Two particularly good surveys of this subject are:

    a. "Spring Symposia", (NRL Report 6972) Naval Research Laboratory, Washington, D. C., 1969, vol. 2.

    b. J. K. Hulm, et. al., Progress in Low Temperature Physics (C. J. Gorter, ed.) North-Holland, Amsterdam, 1970, vol. 6, p. 205.

9. Reviews of possible mechanisms and enhancement techniques are found in Ref. (1a) and (8b). See also J. W. Garland, "Spring Symposia", (NRL Report 7023) vol. 1, p. 32; V. L. Ginzburg, Sov. Phys.-Uspekhi 13, 335 (1970); B. Matthias, et. al., Phys. Rev. Letters 27, 245 (1971).

10. W. L. McMillan, Phys. Rev. 167, 331 (1968). See also W. Klose and P. Hertel, Z. Physik 239, 331 (1970) and R. Benda, Phys. Stat. Sol. 38, 451 (1970).

11. P. E. Seiden, Phys. Rev. 168, 403 (1968); F. Holtzberg, et. al., Phys. Rev. 168, 408 (1968); R. J. Gambino and P. E. Seiden, Phys. Rev. B2, 3571 (1970).

12. P. E. Seiden, "Spring Symposia", (NRL Report 6972) vol. 2, p. 45.

13. R. C. Dynes, Phys. Rev. B2, 645 (1970); G. Deutscher, et. al., Phys. Letters 35A, 265 (1971); J. J. Hanak, et. al., Phys. Letters 30A, 201 (1969).

14. J. J. Hopfield, Phys. Rev. 186, 443 (1969); J. W. Garland and K. H. Benneman, unpublished; S. Barisic, et. al., Phys. Rev. Letters 25, 919 (1970).

15. J. W. Garland, Phys. Rev. 153, 460 (1967).

16. E. E. Havinga, et. al., J. Phys. Chem. Solids 31, 2653 (1970) and E. E. Havinga and M. H. van Maaren, Proceedings of the 12th International Conference on Low Temperature Physics (Eizo Kanda, ed.), Academic Press of Japan, Tokyo, 1971, p. 355.

17. Leon N. Cooper and Birger Stölan, Phys. Rev. B4, 863 (1971).

18. Sang Boo Nam, Phys. Rev. B3, 2946 (1971).

19. W. W. Webb, J. Appl. Phys. 42, 107 (1971).

20. Y. Simon and P. Thorel, Phys. Letters 35A, 450 (1971).

21. Y. B. Kim, Proceedings of the 12th International Conference on Low Temperature Physics (Eizo Kanda, ed.), Academic Press of Japan, Tokyo, 1971, p. 231, and Eugene W. Urban, J. Appl. Phys. 42, 115 (1971).

22. N. F. Berk and J. R. Schrieffer, Proceedings of the 10th International Conference on Low Temperature Physics, Moscow 1966, Venete, Moscow, 1967, vol. 2A, p. 150.

23. See for example, Roger W. Cohen, et. al., Proceedings of the 3rd IMR Symposium on the Electronic Density of States (NBS Spec. Pub. 323), National Bureau of Standards, Washington, D. C., 1970.

24. See, e. g., J. Labbé, Proceedings of the 10th International Conference on Low Temperature Physics, Moscow 1966, Venete, Moscow, 1967, vol. 3, p. 264.

25. P. W. Anderson and E. J. Blount, Phys. Rev. Letters 14, 217 (1965).

26. G. Shirane and J. D. Axe, to be published.

27. W. Klose and H. Schuster, Z. Physik 241, 348 (1971).

28. W. Dieterich and H. Schuster, Phys. Letters 35A, 48 (1971).

29. J. J. Hauser, et. al., Phys. Rev. 151, 296 (1966); V. Hoffstein and R. W. Cohen, Phys. Letters 29A, 603 (1969); M. Weger, Sol. State Comm. 9, 107 (1971).

30. M. Weger, J. Phys. C 4, L188 (1971).

31. R. H. Willens, et. al., Sol. State Comm. 7, 837 (1969).

32. Pierre Spitzli, Chaleur Specifique D'Alliages de Structure A-15, (unpublished thesis, Université de Genève, 1970).

33. H. Fröhlich, J. Phys. C 1, 544 (1968); Phys. Letters 26A, 169 (1968); Phys. Letters 35A, 325 (1971). A. Rothwarf, Phys. Rev. B2, 3560 (1970). See also H. Büttner and E. Gerlach, Sol. State Comm. 9, 1407 (1971) for a different application.

34. A. I. Golovashkin, et. al., Sov. Phys.-JETP 30, 44 (1970) and Sov. Phys.-JETP 32, 1064 (1971).

35. L. P. Gor'kov, Sov. Phys.-JETP 7, 505 (1958).

36. Vinay Ambegaokar, Superconductivity (R. D. Parks, ed.), Marcel Dekker, Inc., New York, 1969, vol. 1, p. 259.

37. B. D. Josephson, Superconductivity (R. D. Parks, ed.), Marcel Dekker, Inc., New York, 1969, vol. 1, p. 423.

38. Elihu Abrahams and T. Tsuneto, Phys. Rev. 152, 416 (1966); M. C. Leung, Can. J. Phys. 48, 598 (1970); L. P. Gor'kov and G. M. Eliashberg, Sov. Phys.-JETP 27, 328 (1968).

39. G. Lucas and M. J. Stephens, Phys. Rev. 154, 349 (1967); Albert Schmid, Phys. Kond. Materie 5, 302 (1966); James Woo and Elihu Abrahams, Phys. Rev. 169, 407 (1968); T. J. Rieger, Phys. Rev. B3, 2253 (1970); A. Houghton and E. Maki, Phys. Rev. B3, 1625 (1971).

40. Wayne Hudson, The Mixed State of Superconductors (NASA SP-240), National Aeronautics and Space Administration, Washington, D. C., 1970.

41. P. W. Anderson, Comments on Sol. State Phys. 1, 31 (1968) and Sol. State Phys. 1, 190 (1969); Phys. Rev. B3, 153 (1971); to be published. Also, A. Griffin, Superconductivity (P. R. Wallace, ed.), Gordon and Breach, New York, 1969, vol. 2, p. 577.

42. P. Fulde, Phys. Rev. 137, A783 (1965) and Tunneling Phenomena in Solids (E. Burstein and S. Lundqvist, eds.), Plenum Press, New York, 1969, p. 427. See also Sang Boo Nam, Phys. Rev. 156, 470 (1967) and Phys. Rev. 156, 487 (1967).

43. David Cohen, et. al., Appl. Phys. Letters 16, 278 (1970) and J. E. Zimmerman and N. V. Frederick, Appl. Phys. Letters 19, 16 (1971).

Part VI
Appendices

APPENDIX 1

PROGRAM FOR

THE SCIENCE AND TECHNOLOGY OF SUPERCONDUCTIVITY

An Intensive Course Co-Sponsored by
The School for Summer and Continuing Education
and the
Department of Physics
Georgetown University

August 13-26, 1971
Reiss Building
Georgetown University
Washington, D. C.

Directors: William D. Gregory
Wesley N. Mathews Jr.
Edgar A. Edelsack

Date and Time	Video Tape Reference No.	Topic and Speaker
Thursday, August 12		
3:00 - 6:00 p.m.		Registration
Friday, August 13		SESSION I
9:00-9:30 a.m.		Orientation W. D. Gregory, E. A. Edelsack, W.N. Mathews Jr.
9:30-10:30 a.m.	S1	Fundamentals of Superconductivity E. A. Edelsack
10:30-11:00 a.m.		Coffee
11:00 a.m.-Noon	S2	Phenomenological Theories of Superconductivity, I W. D. Gregory
Noon-2:00 p.m.		Lunch

	SESSION II	
2:00-3:00 p.m.	S3	Phenomenological Theories of Superconductivity, II W. D. Gregory
3:15-4:15 p.m.	S4	Elements of the Theory of Superconductivity, I W. N. Mathews Jr.
4:15-5:30 p.m.		Consultation, Individual Discussions
Saturday, August 14	SESSION III	
1:00-2:00 p.m.	S5	Elements of the Theory of Superconductivity, II W. N. Mathews Jr.
2:00-2:30 p.m.		Coffee
2:30-3:30 p.m.	S6	Elements of the Theory of Superconductivity, III W. N. Mathews Jr.
Monday, August 16	SESSION IV	
9:00-10:15 a.m.	S7	Elements of the Theory of Superconductivity, IV W. N. Mathews Jr.
10:15-10:45 a.m.		Coffee
10:45 a.m.-noon	S8	Superconductivity in Very Pure Metals W. D. Gregory
Noon-2:00 p.m.		Lunch
	SESSION V	
2:00-3:15 p.m.	A76 A77 B29	Laboratory Demonstrations W. D. Gregory
3:15-3:45 p.m.		Coffee
3:45-5:00 p.m.		Laboratory Demonstrations W. D. Gregory

APPENDIX 1   759

Tuesday, August 17   SESSION VI

9:00-10:15 a.m.   S9   Elements of the Theory of
                                      Superconductivity, V
                                      W. N. Mathews Jr.

10:15-10:45 a.m.   Coffee

10:45 a.m.-noon   S11   Enhancement of Power Output
                                      from Superconducting
                                      Radiation Devices
                                      W. D. Gregory

Noon-2:00 p.m.   Lunch

                      SESSION VII

2:00-3:15 p.m.   S12   Superconductors in
                                      Thermometry
                                      J. F. Schooley

3:15-3:45 p.m.   Coffee

3:45-5:00 p.m.   S10   Elements of the Theory of
                                      Superconductivity, VI
                                      W. N. Mathews Jr.

Wednesday, August 18

8:30 a.m.-5:30 p.m.   Visits to local laboratories:
                                      1) Naval Research Laboratory
                                      2) University of Maryland
                                      Low Temperature Group
                                      3) National Bureau of
                                      Standards
                                      Assemble at Healy Circle at
                                      8:30 a.m.

Thursday, August 19   SESSION VIII

8:30-10:00 a.m.   S12   Enhancement Effects
                                      J. F. Schooley

10:00-10:30 a.m.   Coffee

10:30 a.m.-Noon   S13   Enhancements Effects:
                                      Theory
                                      C. S. Koonce

Noon-2:00 p.m.   Lunch

SESSION IX

2:00-3:30 p.m.	S14	The Use of Superconductivity for Metrology T. F. Finnegan
3:30-4:00 p.m.		Coffee
4:00-5:30 p.m.	S15	Superconducting Intermetallic Compounds - The A15 Story R. A. Hein

Friday, August 20 — SESSION X

8:30-10:00 a.m.	S16 S16.1	Physics of Superconducting Devices B. Deaver
10:00-10:30 a.m.		Coffee
10:30 a.m.-noon	S17	The Josephson Effect F. Bloch
Noon-2:00 p.m.		Lunch

SESSION XI

2:00-3:30 p.m.	S18	Refrigeration for Superconducting Devices R. W. Stuart
3:30-4:00 p.m.		Coffee
4:00-5:30 p.m.	S20.1 S20	Theory of Superconducting Semiconductors C. S. Koonce

Saturday, August 21 — SESSION XII

9:00 a.m.-5:00 p.m.		Displays of Commercial Superconducting and Cryogenic Devices Literature Display Informal Discussions Coffee

APPENDIX 1

Monday, August 23	SESSION XIII	
8:30-10:00 a.m.	S21 S21.1	The Metallurgy of Superconductors R. M. Rose
10:00-10:30 a.m.		Coffee
10:30 a.m.-noon	S22 S22.1	Superconductive Computer Devices J. Matisoo
Noon-2:00 p.m.		Lunch
	SESSION XIV	
2:00-3:30 p.m.	S23 S23.1	Superconducting Power Transmission R. W. Meyerhoff (delivered by D. A. Haid)
3:30-4:00 p.m.		Coffee
4:00-5:30 p.m.	S24 S24.1	Superconducting Coils Z. J. J. Stekley
7:00-9:00 p.m.	S29 S29.1	Application of Superconductivity in Thermonuclear Fusion Research A. Martinelli
Tuesday, August 24	SESSION XV	
8:30-10:00 a.m.	S25 S25.1	Application of Superconductors to Motors and Generators J. L. Smith, Jr.
10:00-10:30 a.m.		Coffee
10:30 a.m.-noon	S26 S26.1	Time Dependent Superconductivity D. J. Scalapino
Noon-2:00 p.m.		Lunch

SESSION XVI

2:00-3:30 p.m.	S27	Superconducting Accelerators M. McAshan
3:30-4:00 p.m.		Coffee
4:00-5:30 p.m.	S28 S28.1	Electric and Magnetic Shielding with Superconductors Blas Cabrera and W. O. Hamilton (delivered by W. O. Hamilton)

Wednesday, August 25   SESSION XVII

8:30-10:00 a.m.	S30 S30.1	Tc's -- The High and Low of It B. T. Matthias
10:00-10:30 a.m.		Coffee
10:30 a.m.-noon	S31	Levitation Devices for Transportation H. T. Coffey
Noon-2:00 p.m.		Lunch

SESSION XVIII

2:00-3:00 p.m.	S32	Panel Discussion: The Scientific, Technological, and Economic Implications of Advances in Superconductivity
3:00-3:30 p.m.		Coffee
3:30-5:00 p.m.	S32.1	Panel Discussion (Contd.)
5:30-7:00 p.m.		Refreshments
7:00 p.m.		Banquet

Thursday, August 26   SESSION XIX

8:30-10:00 a.m.	S33 S33.1	Millimeter and Submillimeter Detectors and Devices S. Shapiro

APPENDIX 1

10:00-10:30 a.m.   Coffee

10:30 a.m.-noon   S34   Magnetometers and Inter-
                      S34.1   ference Devices
                               W. W. Webb

Noon-2:00 p.m.   Lunch

                      SESSION XX

2:00-3:30 p.m.   S35   The Technological Implica-
                                  tions of Superconductivity
                                  in the Next Decade
                                  D. Langenberg

3:30-4:00 p.m.   Coffee

4:00-5:30 p.m.   S36   A Summary of the Course
                                  J. L. Bostock

APPENDIX 2

THE SCIENCE AND TECHNOLOGY OF SUPERCONDUCTIVITY

Invited Speakers

Professor F. Bloch
Stanford University
Department of Physics
Stanford, California 94305

Dr. Judith L. Bostock
Institut für Theoretische
Festkörperphysik
Universität des Saarlandes
66 Saarbrücken 15
Germany

Dr. H. T. Coffey
Stanford Research Institute
333 Ravenswood Avenue
Menlo Park, California 94025

Professor B. S. Deaver, Jr.
Department of Physics
University of Virginia
Charlottesville, Va. 22903

Mr. E. A. Edelsack
3510 Hamlet Place
Chevy Chase, Maryland 20015

Dr. T. F. Finnegan
National Bureau of Standards
Gaithersburg, Maryland 20234

Professor W. D. Gregory
Department of Physics
Georgetown University
Washington, D.C. 20007

Mr. David A. Haid
Union Carbide Corporation
Tarrytown, New York 10591

Professor W. O. Hamilton
Department of Physics
Louisiana State University
Baton, Rouge, Louisiana 70803

Dr. R. A. Hein
Naval Research Laboratory
Washington, D. C. 20390

Dr. C. S. Koonce
National Bureau of Standards
Gaithersburg, Maryland 20234

Professor D. Langenberg
Department of Physics
University of Pennsylvania
Philadelphia, Pennsylvania 10104

Dr. Michael S. McAshan
High Energy Physics Laboratory
Stanford University
Stanford, California 94306

Dr. Alberto Martinelli
Institut für Plasmaphysik
8046 Garching bei München
West Germany

Professor W. N. Mathews Jr.
Department of Physics
Georgetown University
Washington, D. C. 20007

Dr. J. Matisoo
IBM Watson Research Center
Yorktown Heights, N. Y. 10598

Professor B. T. Matthias
Department of Physics
University of California
La Jolla, California 92037

Professor R. M. Rose
Department of Metallurgy
Massachusetts Institute of Technology
Cambridge, Massachusetts 02139

Professor D. J. Scalapino
Department of Physics
University of California
Santa Barbara, California 93106

Dr. J. F. Schooley
National Bureau of Standards
Gaithersburg, Maryland 20234

Professor S. Shapiro
Department of Electrical Engineering
University of Rochester
Rochester, New York 14627

Professor J. L. Smith, Jr.
Department of Electrical Engineering
Massachusetts Institute of Technology
Cambridge, Massachusetts 02139

Dr. Z. J. J. Stekly
Magnetic Corporation of America
67 Rodgers Street
Cambridge, Massachusetts 02142

Mr. Robert W. Stuart
Cryogenic Technology, Inc.
Waltham, Massachusetts 02154

Professor W. W. Webb
Department of Physics
Cornell University
Ithaca, New York 14850

APPENDIX 3

THE SCIENCE AND TECHNOLOGY OF SUPERCONDUCTIVITY

Participant List

Friedrich W. Ackermann
30 Larchwood Road
Wyomissing, Pennsylvania 19610

G. Douglas Baker
Physics Department
Louisiana State University
Baton Rouge, Louisiana 70803

James W. Baker
Texas A&M University
College Station, Texas 77843

Mohamed Behravesh
2800 Quebec St., N. W. #515
Washington, D. C. 20008

Dr. James Beitchman
U.S. Army Foreign Science
Technology Center
220 Seventh St., N.E.
Charlottesville, Virginia 22901

Thomas Lee Berger
27 Pendleton Rd.
Fredricksburg, Virginia 22401

Gerald V. Blessing
1806 Metzerott Road, #A4
Adelphi, Maryland 20783

Larry H. Capots
3722 R Street, N.W.
Washington, D. C. 20007

Peter F. Carcia
The Pennsylvania State Univ.
Materials Research Laboratory
University Park, Pennsylvania

Allen Carroll
Department of Physics
Clark Hall
Cornell University
Ithaca, New York 14850

Dr. P. J. Carroll
Naval Ship Research &
Development Laboratory
Code P133
Panama City, Florida 32401

Clifford Chapman
7521 McWhorter Place, Apt. T-3
Annandale, Virginia 22003

Woontong Nathan Cheung
Department of Metallurgy
Rm. 8-411
Massachusetts Institute
of Technology
Cambridge, Massachusetts 02139

Nicholas C. Cirillo
1917 Kennedy Drive
McLean, Virginia 22101

Dr. Badri Das
Naval Research Laboratory
Metallurgy Division
Code 6350
Washington, D. C. 20390

Robert J. Dinger
Code 604
Naval Weapons Center
China Lake, California 93555

George Durako
2302 4th Street
Easton, Pennsylvania 18402

K. A. Evans
5002 Gadson Drive
Fairfax, Virginia 22030

Dr. Stephen J. Fonash
231-B Sackett Building
Pennsylvania State University
University Park, Penna. 16802

Thomas L. Francavilla
3400 Lorring Drive, #302
Washington, D. C. 20028

Myron H. Frommer
Room 8-409
Massachusetts Institute
of Technology
Cambridge, Massachusetts 02139

Dr. Martin R. Gaerttner
Department of Chemistry
Renssalaer Polytechnic Institute
Troy, New York

Charles S. Gault
Physics Department
American University
Massachusetts & Nebraska Ave.
Washington, D. C. 20016

James P. George
809 West Broad Street #417
Falls Church, Virginia 22046

Andrew J. Grekas
402 Watkins Trail
Annandale, Virginia 22003

Ronald J. Gripshover
RD 2 Box 93B
King George, Virginia

Paul Hankinson
2136 City Line Road
Bethlehem, Pennsylvania 18017

Mr. Leslie Harner
2136 City Line Road
Bethlehem, Pennsylvania 18017

Stuart Horn
3214 Lothian Road
Fairfax, Virginia 22030

Lawrence D. Jackel
Department of Physics
Clark Hall
Cornell University
Ithaca, New York 14850

Edward W. Johanson
Argonne National Laboratory
D818
9700 S. Cass Avenue
Argonne, Illinois 60439

V. R. Kalvey
1724 17th Street, N. W.
Washington, D. C. 20009

Douglas A. Koop
Kawecki Berylco Industries
P. O. Box 1462
Reading, Pennsylvania 19603

Roger A. Little
Physics Department
Georgia Institute of Technology
Atlanta, Georgia 30332

Dr. Alberto Martinelli
Institut für Plasmaphysik
8046 Garching bei München
West Germany

Robert J. Mattauch
Department of EE
University of Virginia 22903

Dr. Russell A. Meussner
Naval Research Laboratory
Metallurgy Division
Code 6350
Washington, D. C. 20007

John Miller
Department of Physics
University of Virginia
Charlottesville, Virginia 22103

APPENDIX 3

Richard A. Moyle
2438 39th Place, N. W.
Washington, D. C. 20007

Rassmidara Navani
1445 44th Street, N. W.
Washington, D. C. 20007

Henry R. Odom
P. O. Box 49
Dahlgren, Virginia

Morris A. Olson
2 617 Kirkwood Place
Hyattsville, Maryland

Rambhai M. Patel
c/o Mr. P. K. Mehta
4205 Military Road
Washington, D. C.

Louis A. Pendrys
102 Lattimore Road
Rochester, New York 14620

Dr. Raymond S. Ramshaw
Dept. of Electrical Engineering
University of Waterloo
Waterloo, Ontario, Canada

Dr. R. Allen Reese
Union Carbide Corporation
Saw Mill River Road at Rt. 100C
Tarrytown, New York 10591

Dominic J. Repici
1855 Old Meadow Road
McLean, Virginia 22101

Raymond Roberge
Hydro-Quebec Institute of Research
Varennes, Quebec, Canada

Mr. Henry A. Scheetz
5001 Caryn Court #201
Alexandria, Virginia 22312

Mr. Carl Schueler
Physics Department
Louisiana State University
Baton Rouge, Louisiana 70803

Heinz Schuster
Institut für theoretische
Festkörperphysik
Universität des Saarlandes
66 Saarbrücken 15
Germany

William D. Smith
Carborundum Co., R&D Division
P.O. 337
Niagara Falls, New York 14302

Charles V. Stancampiano
803 University Park
Rochester, New York 14620

Philip A. Struder
10313 Ridgemoor Drive
Silver Spring, Maryland

Dr. M. H. Talaat
3829 Hamilton Street
Hyattsville, Maryland

Ariel A. Valladares
Instituto de Fisica
Apartado Postal 20-364
Mexico 20, D. F.
Mexico

Dr. Ronald G. Vardiman
Naval Research Laboratory
Metallurgy Division
Code 6350
Washington, D. C. 20390

A. A. Wolf
RFD 3
Box 327
Annapolis, Maryland 21403

Andrew Wolski
Department of Science
Universite de Moncton
Moncton, Nouveau Brunswick, Canada

Dr. W. M. Wynn
Naval Ship Research and Development
Laboratory
Code P133
Panama City, Florida 32401

APPENDIX 4

PROBLEMS

The following are problems on some of the material covered in the course. Some of these problems were assigned as homework for those taking the course for credit.

Part A

Problem (J. Bostock)

Tantalum has a critical temperature of 4.4 K in zero magnetic field and a critical field at absolute zero of 830 Oe. If a Ta wire of diameter 100 microns is at 4K, what is the approximate value for the maximum supercurrent (in amps) that can be carried by the wire in the absence of an applied field? (Hint: Use Silsbee's Rule.) What is the approximate critical current density (amps/cm^2) in this wire? How do the current density and the current depend on wire diameter?

Problem on "Enhancement Effects: Theory" (C. S. Koonce)

Consider the approximate equation for the transition temperature including renormalization effects, but neglecting the Coulomb pseudopotential, μ^*,

$$T_c = \tilde{\omega} \exp(-(1+\lambda)/\lambda),$$

where $\tilde{\omega}$ is taken to be an <u>observed</u> phonon frequency. In the talk on enhancement effects we saw that this frequency could be expressed in terms of the bare frequency, ω_λ, and the electron-phonon coupling, g, as

$$\tilde{\omega}^2 = \omega_\lambda^2 - \frac{2g^2 \omega_\lambda}{V_c} \frac{\varkappa_e - 1}{\varkappa_e},$$

where V_c is the bare Coulomb interaction and $\varkappa_e$ is the electronic contribution to the dielectric function. If we approximate λ by $\lambda = \frac{Kg^2}{\tilde{\omega}^2}$, where K is a constant, we may express the transition temperature in terms of the electron-phonon coupling and the bare phonon frequencies. Suppose that by alloying or black magic you are able to choose the electron-phonon coupling, g, to be whatever you need to produce a high transition temperature superconductor.

A. What value of g do you choose? Express your result, if finite, in terms of K, ω_λ, V_c and $\varkappa_e$.

B. What happens when $g^2 = g_u^2 = V_c \omega_\lambda \varkappa_e /2(\varkappa_e - 1)$?

C. Make a rough plot of T_c vs. g.

D. If we express the result of part A as g_M, make a rough plot of Kg_M^2/ω_λ^2 vs. Kg_u^2/ω_λ^2.

E. If $x = g_M^2/g_u^2$, express the maximum attainable ratio of the transition temperature to the bare phonon frequency as a function of x.

Problem on "Time Dependent Superconductivity" (D. J. Scalapino)

When Rieger, Mercereau and I submitted some of these ideas in a manuscript to Physical Review Letters it was rejected. Here is the referee's report:

> "I am rather doubtful that the phenomenological equations derived in this paper are correct. They contain the unphysical feature that if a process takes place in which pairs are broken up into quasiparticles or vice versa, a charge density is developed. The strong Coulomb forces within a metal would prevent such a charge from forming. It would be expected that the chemical potential would adjust itself in such a way that $\delta\rho = 0$. A similar problem arises in determining the Bernoulli potential in a superconductor and it is important to take this condition into account. See G. Rickayzen, Journal of Physics C2, 1334 (1969). A realistic calculation would require the solution of the Gorkov equations with the extra condition $\delta\rho = 0$."

How would you answer it?

Hint: In this lecture we have neglected the Coulomb energy in Eq. (2.1) and subsequently in the basic dynamics, Eq. (2.8). This will only make sense if the Coulomb energy associated with a charge density buildup arising from a finite $\nabla \cdot \vec{j}_s$ is small compared with the basic superconducting energies. You can estimate $\delta\rho$ by taking a sphere of radius ξ and assuming that a critical current $\vec{j}_s$ flows uniformly out over this sphere. Then compute the Coulomb energy of a sphere containing a uniform $\delta\rho$ of this size and comparing that with the superconducting condensation energy. You will find that the Coulomb energy is much less than the superconducting condensation energy.

APPENDIX 4

Problem on "T_c's -- The High and Low of It" (B. T. Matthias)

Explain why the superconducting transition temperature is so drastically different in these closely related pairs of compounds:

a) ZrC	not superconducting	NbC 12.5 K
b) Nb_3Sb	not superconducting	Nb_3Sn 18 K
c) V_cAu	2 K	Nb_3Au 12 K

Problem on "Superconducting Coils" (Z. J. J. Stekly)

The individual strand size in a composite conductor strongly influences the conductor performance. Assume that the superconductor in the critical state exposed to a constant magnetic field can be modeled locally by:

1. A critical current density that decreases linearly with increasing temperature (This implies that the flux flow resistivity is large and can be neglected.):

$$j_s = j_c \left[1 - \left(\frac{T - T_b}{T_c - T_b} \right) \right].$$

Here j_c = critical current density at T_b in the presence of a magnetic field and T_c = critical temperature in the presence of magnetic field.

2. Constant thermal conductivity.

(a) Derive the expression for the <u>steady state</u> temperature distribution and the total current in an <u>individual</u> strand of superconductor as a function of electric fields along the strand for a round strand of radius r_w and for a flattened strand of large aspect ratio of thickness $2 r_w$. The strand is in a well cooled matrix of high thermal conductivity so that its surface is maintained at T_b.

(b) Expand the exact expressions for the strand current in terms of series to obtain the behavior near $E = 0$.

(c) Using only the first term in E, arrive at a critical strand size for steady state stability of the strand current when it is at its maximum current carrying capacity ($E \to 0$), for a conductor having a matrix cross section A_m, superconductor cross section A_s, and electrical resistivity of the matrix equal to ρ.

Part B

1. (a) Show that
$$N(0)V_{eff} = -\frac{1}{2 + 6.71\,\alpha\,a_o\,(v_o Z)^{-1/3}} \cdot \frac{1}{\alpha'\,(T_D/T_c)^2 - 1},$$

where

$$\alpha = (k/k_D)^2, \quad \alpha' = \left(\frac{\omega_{\vec{k}}}{\epsilon_{\vec{p}+\vec{k}} - \epsilon_{\vec{p}}}\right)^2 \left(\frac{T_c}{T_p}\right)^2, \quad a_o = \hbar^2/me^2,$$

v_o is the atomic volume, k_D is the Debye wave number, and T_D is the Debye temperature.

(b) Taking $\alpha = 3/5$, estimate the value of α' which will give the experimental value of $N(0)V_{eff} = -0.175$ for Aℓ.

---

2. Prove that the spatial Fourier transform of the screened Coulomb interaction,

$$V_c(\vec{x}) = e^2\,\frac{e^{-k_s|\vec{x}|}}{|\vec{x}|}, \quad \text{is given by} \quad V_c(\vec{k}) = \frac{4\pi e^2}{(k^2 + k_s^2)V_o}.$$

---

3. Using the ideas and approximations discussed in our treatment of "Cooper's problem", show that for a pure material the mean square radius of a Cooper pair is approximately $\frac{2}{3}\frac{\hbar v_F}{|E|}$, where E is the energy of the bound state and v_F is the average Fermi speed.

---

4. Consider the wave function of Cooper's problem. Assume that: (1) the material is pure; (2) the net wave vector of the pair is zero; (3) $\mathcal{E}_{\vec{k}} = \hbar^2 k^2/2m - \mathcal{E}_F$. Show that for large $R \equiv |\vec{x}_1 - \vec{x}_2|$ the wave function drops off like $1/R^2$.

---

5. Consider Cooper's problem for a pure material subject to the restriction that the pair have net wave vector $\vec{Q}$. Show that for small Q the bound state energy is given by

$$E_{\vec{Q}} \simeq E_o + \hbar v_F Q/2.$$

APPENDIX 4

6. Consider the pair destruction and creation operators,

$$b_{\vec{k}} = c_{-\vec{k}\downarrow} c_{\vec{k}\uparrow}, \quad b_{\vec{k}}^\dagger = c_{\vec{k}\uparrow}^\dagger c_{-\vec{k}\downarrow}^\dagger.$$

Work out the following commutators and anticommutators:

$$[b_{\vec{k}}, b_{\vec{k}'}]; [b_{\vec{k}}^\dagger, b_{\vec{k}'}^\dagger]; [b_{\vec{k}}, b_{\vec{k}'}^\dagger]; \{b_{\vec{k}}, b_{\vec{k}'}\}; \{b_{\vec{k}}^\dagger, b_{\vec{k}'}^\dagger\}; \{b_{\vec{k}}, b_{\vec{k}'}^\dagger\}.$$

Here

$$[A,B] \equiv AB - BA, \{A,B\} \equiv AB + BA.$$

Interpret your results.

---

7. Show that the equal time anticommutation rules,

$$\{c_\alpha, c_{\alpha'}\} = \{c_\alpha^\dagger, c_{\alpha'}^\dagger\} = 0, \{c_\alpha, c_{\alpha'}^\dagger\} = \delta_{\alpha,\alpha'},$$

imply that the wave field operators,

$$\psi_s(\vec{x}, t) = \sum_\alpha \phi_\alpha(\vec{x}, s) c_\alpha(t), \quad \psi_s^\dagger(\vec{x}, t) = \sum_\alpha \phi_\alpha(\vec{x}, s)^* c_\alpha^\dagger(t),$$

obey the following anticommutation rules:

$$\{\psi_s(\vec{x}, t), \psi_{s'}(\vec{x}', t)\} = \{\psi_s^\dagger(\vec{x}, t), \psi_{s'}^\dagger(\vec{x}', t)\} = 0,$$

$$\{\psi_s(\vec{x}, t), \psi_{s'}^\dagger(\vec{x}', t)\} = \delta_{s,s'} \delta(\vec{x} - \vec{x}').$$

---

8. Show that the Hamiltonian

$$H = \sum_{\vec{p},\sigma} \epsilon_{\vec{p}} c_{\vec{p}\sigma}^\dagger c_{\vec{p}\sigma} + \frac{1}{2} \sum_{\substack{\vec{p},\vec{p}',\vec{k} \\ \sigma,\sigma'}} V_{\text{eff}}(\vec{p} + \vec{k}, \vec{p}) c_{\vec{p}+\vec{k},\sigma}^\dagger c_{\vec{p}'-\vec{k},\sigma'}^\dagger c_{\vec{p}'\sigma'} c_{\vec{p}\sigma}$$

may also be written as

$$H = \sum_s \int d^3x \, \psi_s^\dagger(\vec{x}, t) \mathcal{H}_e(x, \frac{\hbar}{i}\nabla) \psi_s(\vec{x}, t) +$$

$$\frac{1}{2} \sum_{s,s'} \int d^3x \int d^3x' \, \psi_s^\dagger(\vec{x}, t) \psi_{s'}^\dagger(\vec{x}',t) V_{\text{eff}}(\vec{x}, \vec{x}') \psi_{s'}(\vec{x}', t) \psi_s(\vec{x}, t).$$

Also show that the number operator, $N = \sum_{\vec{p},\sigma} \int c_{\vec{p}\sigma}^\dagger c_{\vec{p}\sigma}$, may be written as

$$N = \sum_s \int d^3 x\, \psi_s^\dagger(\vec{x},\, t)\, \psi_s(\vec{x},\, t).$$

Here the wave field operators are given by

$$\psi_s(\vec{x},\, t) = \sum_{\vec{p},\sigma} \phi_{\vec{p}\sigma}(\vec{x},\, s)\, c_{\vec{p}\sigma}(t)\,,$$

and

$$\psi_s^\dagger(\vec{x},\, t) = \sum_{\vec{p},\sigma} \phi^*_{\vec{p}\sigma}(\vec{x},\, s)\, c^\dagger_{\vec{p}\sigma}(t)\,,$$

and

$$\mathcal{H}_e\!\left(\vec{x},\, \frac{\hbar}{i}\nabla\right)\phi_{\vec{p}\sigma}(\vec{x},\, s) = \epsilon_{\vec{p}\sigma}\, \phi_{\vec{p}\sigma}(\vec{x},\, s).$$

---

9. The BCS choice for the state vector of the superconducting ground state is

$$\psi_s = \prod_n (u_n + v_n b_n^\dagger)\Phi_0,$$

where Φ_0 is the vacuum state and $b_n = c^\dagger_{n\uparrow} c_{\bar{n}\downarrow}$. Show that normalization of ψ_s is most simply guaranteed by $|u_n|^2 + |v_n|^2 = 1$. Interpret $|u_n|^2$ and $|v_n|^2$.

---

10. We have seen that for a pure material the BCS reduced Hamiltonian can be written as

$$H_{\text{red}} = \sum_{\vec{p},\sigma} \epsilon_{\vec{p}}\, n_{\vec{p}\sigma} - 2 \sum_{\substack{\vec{p} \\ p \leqslant p_F}} \epsilon_{\vec{p}} + \sum_{\vec{p} \neq \vec{p}'} V_{\text{eff}}(\vec{p},\vec{p}')\, c^\dagger_{\vec{p}'\uparrow} c^\dagger_{-\vec{p}'\downarrow} c_{-\vec{p}\downarrow} c_{\vec{p}\uparrow}.$$

We have also seen that the BCS choice for the state vector of the superconducting ground state can be written as

$$\psi_s = \prod_{\vec{k}} (u_{\vec{k}} + v_{\vec{k}} b_{\vec{k}}^\dagger)\Phi_0,$$

with

$$|u_{\vec{k}}|^2 + |v_{\vec{k}}|^2 = 1,\quad b_{\vec{k}}^\dagger = c^\dagger_{\vec{k}\uparrow} c^\dagger_{-\vec{k}\downarrow},$$

and Φ_0 the vacuum state.

(a) Show that the energy of the superconducting ground state, relative to the energy of the filled zero temperature Fermi sea, is given by

$$W_0 = 2 \sum_{\substack{\vec{k} \\ k>k_F}} \epsilon_{\vec{k}} |v_{\vec{k}}|^2 + 2 \sum_{\substack{\vec{k} \\ k \leq k_F}} |\epsilon_{\vec{k}}| |u_{\vec{k}}|^2 +$$

$$+ \sum_{\vec{k} \neq \vec{k}'} V_{\text{eff}}(\vec{k}',\vec{k}) u_{\vec{k}}^* v_{\vec{k}} u_{\vec{k}'} v_{\vec{k}'}^*.$$

(b) Show that proper minimization of W_0 with respect to $u_{\vec{k}}$ and $v_{\vec{k}}$, taking account of the fact that

$$|u_{\vec{k}}|^2 + |v_{\vec{k}}|^2 = 1,$$

leads to

$$|u_{\vec{k}}|^2 = 1 - |v_{\vec{k}}|^2 = \frac{E_{\vec{k}} + \epsilon_{\vec{k}}}{2E_{\vec{k}}},$$

$$E_{\vec{k}} = (\epsilon_{\vec{k}}^2 + \Delta_{\vec{k}}^2)^{1/2},$$

and $\Delta_{\vec{k}}$ given by

$$\Delta_{\vec{k}} = \sum_{\substack{\vec{k}' \\ \vec{k}' \neq \vec{k}}} |V_{\text{eff}}(\vec{k}'\vec{k})| \frac{\Delta_{\vec{k}'}}{2E_{\vec{k}'}}.$$

(c) Show that with the values of $u_{\vec{k}}$ and $v_{\vec{k}}$ given in part (b),

$$W_0 = 2 \sum_{\vec{k}} \epsilon_{\vec{k}} |v_{\vec{k}}|^2 - 2 \sum_{\substack{\vec{k} \\ k \leq k_F}} \epsilon_{\vec{k}} - \sum_{\vec{k}} \frac{\Delta_{\vec{k}}^2}{2E_{\vec{k}}}.$$

11. Show that with the approximations,

$$|V_{\text{eff}}(\vec{k}',\vec{k})| \doteq V\theta(\omega - |\epsilon_{\vec{k}}|)\,\theta(\omega - |\epsilon_{\vec{k}'}|),$$

and

$$\sum_{k} \to N(0) \int d\epsilon_{\vec{k}}:$$

(a) the gap equation obtained in part (b) of the previous problem leads to

$$\Delta = \omega/\sinh[1/N(0)V];$$

(b) the energy, W_0, reduces to

$$W_0 = -\frac{2N(0)\omega^2}{e^{2/N(0)V}-1}.$$

(c) Obtain the weak-coupling ($N(0)V \ll 1$) and strong-coupling ($N(0)V \gg 1$) limits of these results.

---

12. (a) Consider an excited state which differs from the ground state only in that $(-\vec{k}'\downarrow)$ and $(\vec{k}''\uparrow)$ are definitely occupied and $(\vec{k}'\uparrow)$ and $(-\vec{k}''\downarrow)$ are definitely unoccupied. Show that the energy of this state is

$$W_{\vec{k}',\vec{k}''} = W_0 + E_{\vec{k}'} + E_{\vec{k}''}.$$

(b) The state vector with an "excited pair" in $\vec{k}'$ and ground pairs elsewhere can be written as

$$\psi_s' = \prod_{\substack{\vec{k} \\ \vec{k} \neq \vec{k}'}} (u_{\vec{k}} + v_{\vec{k}} b_{\vec{k}}^\dagger)(u_{\vec{k}'}^* b_{\vec{k}'}^\dagger - v_{\vec{k}'}^*)\Phi_0.$$

Show that this state is orthogonal to the ground state.

(c) Show that the state of part (b) has energy

$$W_{\vec{k}'} = W_0 + 2E_{\vec{k}'}.$$

AUTHOR INDEX

A

Abeles, B.	247,261,428
Abrahams, E.	164,167,183,748,753
Abrikosov, A. A.	20,71,93,99,128,146,289,710
Adkins, C. J.	329,332
Ageev, N. V.	360,371
Alekseevskii, N. E.	282,284,285,288,360,361,362,371
Allen, J. F.	428
Allen, P. B.	384,387,396,400,403
Allgaier, R. S.	375,386
Alphonse, G. A.	616,622
Ambegaokar, V.	98,684,748,753
Ambler, E.	374,386
Amendas, W.	466,470,472,481
Anacker, W.	621,623
Anderson, P. W.	64,70,114,147,157,161,414,417, 427,547,563,631,632,633,640,650, 654,676,687,690,746,749,752,753
Andres, K.	282,288,427
Appleton, A. D.	486,495
Aron, P. R.	346,370
Arrhenius, G.	291
Asada, T.	358,371
Ashkin, M.	426
Aslamazov, L. G.	643,647,652,684,695
Asprey, L. B.	338,369
Atherton, D.	437,457
Averill, R. F.	57,69,143,148,212,253,260
Axe, J. D.	746,752

B

Babiskin, J.	377,387
Bachner, J. F.	348,370
Baker, D.	604
Bakker, J. W.	628,630
Baldus, W.	190
Baratoff, A.	548,563
Bardeen, J.	7,18,25,29,42,57,68,69,71,74, 77,78,88,90,91,94,95,96,122, 125,128,130,135,140,141,143, 144,147,148,149,152,161,287, 292,325,327,330,331,563,626, 629,685,687,692,737,743,744
Barisic, S.	751
Barnes, L. J.	291
Barrett, C. S.	277,288
Barsa, F.	281,288

Pages 1-428 will be found in Volume 1, pages 429-778 in Volume 2.

Bass, I.	603
Batterman, B. W.	277,288
Baumann, F.	421,428
Baym, G.	71,93,98,99,101,117,139,146,147
Beall, W. T.	439,442,447,457
Bean, C. P.	317,325,331
Beasley, M. R.	668,671,675,677
Beck, P. A.	343,370
Becker, J. G.	374,386
Benda, R.	746,751
Bennemann, K. H.	396,403,751
Benz, M. G.	509,536
Berk, N. F.	745,752
Berkl, E.	474,481
Berlincourt, T. G.	297,330,353,354,371
Berruyer, A.	516,537
Berstein, J. T.	457
Bierstedt, P. E.	291
Bishop, J. H.	268,288
Black, W. C.	268,288
Blackburn, J. A.	548,563
Blackford, B. L.	255,257,261
Blaisse, B. S.	445,457
Blank, C.	445,457
Blatt, F. J.	375
Blatt, J. M.	62,70,374,375
Blaugher, R. D.	291,293,330,353,371,414,426,427
Bloch, F.	38,66,149,155,160,161,171,183, 684,739,751
Blount, E. J.	746,752
Blumberg, W. E.	343,370
Boesenberg, E. H.	433,434,456
Bogoliubov, N. N.	115,147
Bohm, D.	77,146
Bol, M.	593,605
Bonera, G.	281,288
Bosio, L.	226,260
Bostock, J.	57,69,142,148,682,685,716
Boughn, S.	604
Boughton, R. I.	423,428
Bozorth, R. M.	32,69
Bragg	414
Bremer, J. W.	611,622
Brennemaun, A. E.	613,614,622
Brewster, P. M.	268,288
Briscoe, C. V.	248,261
Britton, R. B.	498,535
Brown, R. E.	594,605
Bruning, H. A. C. M.	343,370
Bucher, E.	291,412,427

AUTHOR INDEX

Buchler, R.	313,325
Buck, D. A.	607,622
Buckel, W.	414,423,427,428
Buckner, S. A.	647,652
Buehler, E.	313,325,331
Buhrman, R. A.	667,668,675,676
Buravov, L. I.	282,288
Burger, J. P.	299,301,330
Burns, L. L.	616,622
Burton, E. F.	334,338,369
Buttner, H.	144,148,747,753
Byers, N.	149,152,160

C

Cabrera, B.	587
Cadieu, F. J.	291
Cairns, D. N. H.	449,458
Caroli, C.	327,751
Carroll, P. J.	212,260
Carslaw, H. S.	233,260
Carter, J. T.	157,161
Casimir, H. B. G.	13,54,69,625,629,722
Cave, E. C.	449,450,458
Chandrasekhar, B. S.	38,69
Chanin, G.	242,260
Chapnick, I. M.	340,369
Chester, P. F.	503,535
Chu, C. W.	396,403,426
Clarke, J.	552,561,563,564,568,581,584, 585,631,636,651,671,732
Clausis, K.	336,369
Clem, J. R.	218,249,251,252,261
Cline, H. E.	318,319,321,331
Clinton, W. L.	144,148
Clogston, A. M.	298,330,364,365,371,372
Cobble, J. W.	336,341,369
Cochran, J. F.	211,217,218,219,231,236,237, 244,245,248,259,260,261
Cody, G. D.	346,347,370,372
Coffey, H. T.	682,713,714
Cohen, B. M.	203
Cohen, D.	603,605,675,677,699,753
Cohen, M. H.	428,720,721,733
Cohen, M. L.	285,374,377,378,380,384,386, 387,412,421,426,427
Cohen, R. W.	247,261,366,372,428,746,747, 749,750,752
Coles, W. D.	516,536

Colling, D. A.	291
Collins, S. C.	8,190
Collver, M.	272,288
Colwell, J. H.	209,627,630
Compton, V. B.	292,341,370,426
Conkersloot, H. C.	343,370
Cooper, A. S.	336,369,372
Cooper, J. L.	347,370
Cooper, L. N.	7,18,20,68,71,74,78,87,88,90, 91,94,95,96,112,125,126,128, 130,135,141,143,147,149,152, 161,292,330,629,692,737,744,752
Corak, W. S.	625,629
Corenzwit, E. C.	291,292,316,331,336,341,346, 369,370,371,372,426
Cosentina, L. S.	614,622
Courtney, T. H.	311,313,348,370
Cox, J. E.	209,360,371,414,427,736,750
Crane, L. T.	545,563
Crippa, M. L.	281,288
Critchlow, P. R.	321
Crow, J. E.	428
Crowe, J. W.	614,622
Cullen, A. L.	435,457
Cupp, J. D.	571,572,584

D

Daniels, A.	192,203
Danielson, G. C.	417,427
Darnell, F. J.	291
Darsi, D.	343,370
Daunt, J. G.	193,217,260,291,336,338,339, 341,369,375,386
Davies, R. O.	426
Dayem, A. H.	640,651,668,670
Deaver, B. S.	7,15,69,539,563,589,591,592, 593,602,605,682,687,738
Defrain, A.	226,260
deGennes, P. G.	59,70,82,144,147,299,317,325, 327,330,331,548,563
Deis, D. W.	426
DeLaunay, J.	337,338,369
Delile, G.	437,457
Denestein, A.	152,567,568,569,573,578,579, 584,682
DeSorbo, W.	142
Deutscher, G.	751
Devlin, G. E.	56,69

DeWaele, A. T. A. M.	595,605,654,676
Diepers, H.	328,332
Dietrich, W.	746
Dillinger, J. R.	428
DiNardo, A. J.	631,637,650
Dolecek, R. D.	337,338,369
Doll, R.	7,15,38,69,589
Douglass, D. H.	291,421,428,628,630
Dowman, J. E.	254,255,256,257,261,689
Drangeid, K. E.	608,622
Dudley, J. C.	489,490,496
duPre, F. K.	192,203
Dy, K. S.	143,148
Dynes, R. C.	690,742,751
Dziuba, R. F.	577,578,584
Dzyaloshinski, I. E.	71,93,99,146

E

Eagar, T.	316,331
Eck, R. E.	67
Edelsack, E. A.	5,25,603,605,675,677,681,682
	684,685,686,687,688,690,692,
	694,695,696,697,698,699,701
	702,703,704,705,706,707,708,
	709,711,712,713,714,715,716,
	736,737,750
Edwards, D. R.	447,458
Edwards, D. V.	375,386
Ehrenfest, P. E.	47,69
Ehrenfests, T.	69
Eigenbrod, L. K.	439,442,447,457
Eilenberger, G.	144,148,686
Einstein, A.	719,732
Einstein, T. H.	490,496
Eisinger, J.	343,370
Eliashberg, G. M.	167,183,380,381,382,386,387,
	392,396,402,403,748,753
Erdboin, I.	226,260
Essmann, U.	310,331
Evans, D. M.	247,261
Evenson, K. M.	571,572,584

F

Fack, H.	569,573,580,584,647,652
Fairbank, W. M.	7,15,38,69,589,602,603,604,605
Fairbanks, D. F.	509,536
Falge, R. L.	224,260,336,369

Falicov, L. M.	628,630
Falkoff, D.	93,147
Fast, R. F.	516,536
Fermi, E.	168,169
Ferrell, R. A.	298,330,695
Fetter, A. L.	71,93,99,135,136,139,142,146, 317
Feynman, R. P.	70
Field, B. F.	579,585
Filippovich, E. I.	284,288
Finlayson, D. M.	428
Finnegan, T. F.	565,567,568,569,573,578,579, 584,585,647,652,682,699,700
Firsov, Yu. A.	376,387
Fishlock, D.	750
Fitzgerald, R. W.	268,288,291,371
Fiukiger, R.	353,354,371
Foner, A.	316,717
Foner, S.	460,481
Forgacs, R. L.	595,605,667,676
Forshey, R. K.	291
Fowler, R. D.	338,369,691
Fox, G. R.	435,457
Freake, S. M.	329,332
Frederick, N. V.	749,753
Frederiksc, H. P. R.	375,377,386,387,426
Freeman, D. C.	437,439,457
Frenkel, R. B.	569,580,584
Friedel	365,372
Fröhlich, H.	7,16,77,146,334,369,387,625, 629,747,753
Fuchs, K.	245,260
Fulde, P.	749,753
Furdyna, J. K.	301,330

G

Gallop, J. C.	569,579,584
Gambino, R. J.	751
Gamble, F. R.	420,427,428
Gange, R. A.	616,622
Garland, J. W.	396,403,691,743,751
Garwin, R. L.	437,457
Gatos, H. C.	291,347,348,370
Gauster, W. F.	501,535
Gavaler, J. R.	313,331
Geballe, T. H.	266,282,287,288,316,331,341, 346,369,370,371,427
Geiger, A. L.	143,148,247,261

AUTHOR INDEX

Geist, J. M.	188
Geller, S.	341,369,370,375,387,417,427
Gerlach, E.	747,753
Giaever, I.	7,23,209,247,261,428,562,564, 627,630,724
Gibson, J. W.	336,369,375,386
Gierke, G. V.	468,481
Gifford, W. E.	191
Giftleaen	421
Ginsberg, D. M.	242,260
Ginzburg, V. L.	16,58,68,69,70,149,152,153,161, 163,165,167,187,285,289,294, 297,330,676,748,751
Giorgi, A. L.	273,288
Gittleman, J. I.	325,326,331,421,428
Gladstone, G.	397,403,405,426
Glover, R. E. III	209,247,261,414,416,418,420, 421,427,428
Goldschuartz, J. M.	445,457
Goldstein, R.	341,370
Golibersuch, D.	399,403
Gollub, J. P.	675,677
Golovashkin, A. I.	747,753
Gomes deMesquita, A. H.	343,370
Goodkind, J.	660,668,675,676,677
Goodman, B. B.	7,294,297,330,333,334,369,625, 629
Goodman, W. L.	563
Goodstein, J.	732
Goree, W. S.	193,591,592,593,594,721,733
Gor'kov, L. P.	71,112,146,147,167,168,169,183, 294,297,330,748,753
Gorter, C. J.	13,54,69,290,426,428,625,629, 722
Graham, G. M.	427
Graneau, P.	435,457
Green, D. L.	490,496
Greenfield, P.	343,370
Gregory, E.	321
Gregory, W. D.	25,57,69,143,148,185,209,211, 212,218,219,242,244,245,246,247, 249,251,253,254,255,259,260, 261,321,627,629,681,682,689, 703,737,738,750
Greytak, T. J.	291
Grieger, G.	468,481
Griffin, A.	749,753
Grigsby, R.	449,450,458
Grimes, C. C.	631,635,636,637,640,650

Gubser, D. U.	209,217,246,254,261
Gurevich, V. L.	376,387

H

Haden, C. R.	327,331
Hafstrom, J. W.	251,261,328,329,332
Hagner, R.	360,361,371,426
Haid, D. A.	682,707,708,713
Hake, R. R.	291,297,330
Halas, E.	486,495
Halbritter, J.	327,329,331,711
Halloran, J. J.	366,372
Halperin, W. P.	667,675,676
Ham, F. S.	246,261
Hamer, W. J.	568,584
Hamilton, C. A.	209,632,636,643,647,651,652
Hamilton, W. O.	38,587,588,682,701,702
Hammond, R. H.	244,272,288,358,359,371
Hamon, B. V.	575,576,584
Hanak, J. J.	346,347,348,370,421,428,751
Hancox, R.	465,468,479,481,504,535,536
Hannay, N. B.	282,288,427
Hansen, M.	426
Harding, J. T.	581,585,649,652,660,662,668,676
Hardy, G. F.	333,340,341,369
Harland, H. B.	427
Harris, E. P.	626,629
Hart, H. R.	244,254,503,535
Hartsough, L. D.	358,359,371
Hartwig, W. H.	203,327,328,331
Harvey, I. K.	568,584
Hatch, A. M.	486,495
Hauser, J. J.	418,427,747,752
Havinga, E. E.	427,743,752
Heath, F. G.	608,622
Hechler, K.	291
Hein, R. A.	209,224,260,268,270,272,333,336,369,371,375,386,387,412,413,414,427,682,689,692,693,694,700,703,705,706,715,739,745,746
Heiniger, F.	291,353,354,371,412,427
Helf, E.	299,330
Hempstead, C. F.	306,308,331
Hendricks, J. B.	501,535
Hendels, W. H.	632,651
Henning, C. D.	516,520,536

AUTHOR INDEX

Herold, J. S.	426
Hertel, P.	746,751
Hill, D. C.	300,302,303,310,324,331
Hill, H. H.	338,369,691
Hilsch, R.	414,420,427,428
Hitchcock, H. C.	346,370
Hoffstein, V.	747,752
Hogan, W. H.	192,198
Hohenberg, P. C.	142,299,317,330
Holland, L.	247,261
Hollis Hallett, A. C.	427
Holtzberg, F.	375,387,417,427,751
Hood, W. Jr.	675,677
Hopfield, J. J.	399,403,742,751
Hoppe, L. O.	486,495
Horigome, T.	458
Horinchi, T.	358,359,371
Hosler, W. R.	374,375,377,386,387,426
Houghton, A.	748,753
Houston, B. B. Jr.	375,386
Hsu, F. S. L.	313,325,331
Hu, G. R.	299,330
Huang, K.	52,53,69,93,147
Hudson, W.	749,753
Hulbert, J. A.	426
Hull, G. W.	290,346,370,371,375,387,417,427
Hulm, J. K.	291,293,313,330,331,333,340, 341,358,359,369,371,375,386, 391,426,751
Hume-Rothery, W.	406,407,426
Huo, F. S. L.	343,370
Hurwitz, H.	478,481
Hutcherson, J. V.	426

I

Ihara, S.	458
Inoue, K.	292
Irie, F.	308,331
Iwase, Y.	292,460,481

J

Jaccarino, V.	364,365,370,371
Jach, T.	603
Jackson, J. D.	26,68
Jackson, J. E.	248,261
Jacobs, A. E.	144,148,685

Jaeger, J. C.	233,260
Jaklevic, R. C.	143,148,153,161
Janocko, M. A.	313,331
Jansen, H. G.	346,347,370
Jayaraman, A.	375,387
Jennings, L. D.	428
Jensen, M. A.	398,403,405,426
Johnson, R. E.	613,621,622
Johnson, R. W.	190
Jones, C. K.	313,331,426
Josephson, B. D.	7,24,62,70,143,149,150,151,152, 153,154,155,157,158,159,160, 161,163,164,169,171,172,175, 177,181,201,209,563,594,617, 622,631,632,633,640,650,657, 658,659,670,676,719,723,748, 753
Jostram, P. S.	217,260
Junod, A.	353,354,371

K

Kadanoff, L. P.	71,98,99,101,107,121,139,143, 146,147,241,242,244,245,247, 260
Kafka, W.	478,481
Kahn, A. H.	377,387
Kammerer, O. F.	421,428
Kamper, R. A.	628,629,630,632,651,675,677, 721,733,738,750
Kande, E.	426,427,428
Kanter, H.	632,651
Kantrowitz, A. R.	713
Kapitza, P. L.	282,581,585
Karimov, Y. S.	282
Kartsev	486,495
Kazoviskiy	486,495
Keesom, P. H.	217,260
Keister, J. C.	212,244,246,249,251,253,255, 260,261
Khidekel, M. L.	282,288
Kim, Y. B.	143,306,308,330,331,687,744,752
King, H. W.	426
Kinner, H. R.	486,495
Kirschman, R. K.	183
Kirtley, J. L. Jr.	486,490,495
Klaudy, P. A.	437,457
Klose, W.	746,751,752
Knapp, G.	426

AUTHOR INDEX

Kneisel, P.	327,331
Knight, W. D.	244
Kohnlein, D.	423,428
Kohr, J. G.	314,315,316,331
Koonce, C. S.	373,378,380,386,387,389,412, 427,682,686,738,739,740,741, 742,745,751
Korenman, V.	299,330
Kose, V.	556,564,569,573,580,584,647,652
Kramer, L.	291
Krikorian, N. H.	273,288
Krupta, M. C.	273,288
Kümmel, R.	142,144,148,175,183,685
Kunz, W.	291
Kunzler, J. E.	313,325,331,338,343,369,370
Kuper, C. G.	625,629
Kusko, A.	433,456

L

Labbe, J.	364,365,367,372,746,752
Labusch, R.	675,677
LaFleur, W. J.	347,370
Lambe, J. J.	143,148,153,161
Landau, L. D.	16,58,70,78,133,146,147,149, 152,161,163,164,165,167,183, 289,297,300,330,676,737,748
Langenberg, D. N.	67,152,565,567,568,569,573,578, 584,631,633,640,650,682,695, 696,698,702,703,704,714,716, 719,735,739
Langer, J. S.	684
Larkin, A. I.	188,376,387,643,645,647,652, 684,695
Lashmet, P. K.	188
Laverick, C.	504,536
Lawson, A. C.	268
Lawson, J. D.	474,481
Leck, G. W.	616,622
Lee, W. D.	489,496
Leighton, R. B.	70
Leopold, L.	57,69
Lerner, E.	291
Leung, M. C.	144,148,748,753
Levanyuk, A. P.	366,372
Levinstein, H. J.	338,369
Levy, R. H.	516,536
Lewin, J. D.	317,320,321,323,331,536

Lewis, H. W.	57,69
Leyendecker, A. J.	377,387
Lifschitz, E. M.	133,147
Lindsay, J. D. G.	338,369
Little, W. A.	9,81,146,285,544,563,732
Livingston, J. D.	142
London, F.	7,13,14,15,16,32,35,69,141, 147,544,563,589,590,605,722
London, H.	7,13,14,15,16,35,69,722
Long, H. M.	437,439,442,447,457
Longacre, A.	631,640,641,642,647,650,651, 652
Longinotti, L. D.	336,346,369,370,372
Lord, R.	514
Lucas, E. J.	516,536
Lucas, G.	748,758
Luck, D. L.	490,496
Luders, K.	423,428
Lukens, J. E.	672,673,674,675,676,677
Luo, H. L.	291,358,359,371,391,403,426
Lykken, G. I.	143,148,247,261
Lynton, E. A.	241,242,245,260,414

M

Macdonald	736
MacFarlane, J. C.	569,584
MacNab, R. B.	486,495
MacNair, D.	282,288
MacVicar, M. L. A.	247,249,251,254,255,256,257, 271,328,329,330,332,689,704
Maddock, B. J.	501,535
Madey, J.	603
Mahler, W.	343,370
Maita, J. P.	338,343,346,369,370,426
Maix, R.	516,522,537
Maki, E.	142,327,331,748,753
Mapother, D. E.	626,629
March, R. H.	255,257,261
Markowitz, D.	241,242,244,245,247,260
Marshak, H.	629
Martens, H.	328,332
Martinelli, A. P.	459,470,472,478,481,682,712
Mason, P.	703
Mathes, K. N.	445,457
Mathews, W. N. Jr.	27,71,144,148,185,681,682,685, 686,687,737,739
Matisoo, J.	437,457,607,611,617,619,623, 676,682,696,697

Matricon, J.	327,331
Mattheiss, L. F.	372,399,403
Matthias, B. T.	263,266,268,282,291,292,295, 316,333,334,335,336,338,341, 342,353,357,362,366,367,369, 370,372,406,407,410,426,427, 428,682,685,690,691,692,693, 694,706,707,708,709,710,711, 712,714,739,745,751
Mattis, D. C.	42,57,69,140,246,261,327,331
Maxwell, E.	77,146,222,241,260
Mazelsky, R.	375,386
McAshen, M. S.	604,682,709,710,711
McCall, D. M.	428
McCarthy, S.	277,279,288
McConnell,	420
McCumber, D. E.	563,643,647,651
McDermott, R. C.	631,650
McDonald, D. G.	571,572,584,631,647,650
McFee, R.	436,437,457
McInturff, A. D.	322,323,331
McLachlan, D.	242,245,260
McMahon, H. O.	191
McMillan, B.	286,287,384
McMillan, W. L.	338,392,393,395,397,403,426, 683,689,690,691,740,741,751
McNichol, J. J.	613,614,622
McNiff, E. U.	316,331
Meats, R. J.	445,457
Megerle, K.	627,630
Meissner, W.	7,12,13,18,27,334,588,589,591, 593
Melchert, F.	569,573,580,584
Mellors, G. W.	439,440,457
Mendelssohn, K.	32,69
Mercereau, J. E.	65,67,143,153,161,165,173,176, 177,178,179,180,183,563,591, 594,603,605,668,669,670,732
Merriam, M. F.	426
Meservey, R.	143
Meussner, R. A.	209
Meyer, G.	516,522,537
Meyerhoff, R. W.	328,332,433,439,442,447,448,457
Migdal, A. B.	380,387
Miles, J. L.	249,261
Milford, F. J.	375,386
Miller, R. C.	375,386
Minnich, S. H.	435,457
Minors, R. H.	449,458

Mitchell, E. N.   148,247,261
Molokhia, F.   675,677
Monin, F. J.   343,370
Montgomery, D. B.   516,536
Moravesik, M. J.   69
Morin, F. J.   426
Morris, K.   568,569
Morrison, D. D.   308,309,310,324,330,331
Morrison, W. A.   720,721,733
Morse, P. M.   43,69
Moser, S.   421,428
Mota, A. C.   268,288
Mott, N. F.   426
Mueller, F. M.   403,396
Meheim, J.   412,427
Muller, A.   353,354,371
Muller, J.   291,353,354,357,358,363,363,366,371,412,427

N

Näbauer, M.   7,15,38,69,589
Nam, S. B.   744,749,752,753
Nambu, Y.   121,147
Naugle, D. G.   247,261
Neighbor, J. E.   231,260
Nesbitt, L. B.   77
Nethercot, A. H.   242,260
Neuringer, L. J.   325,331
Newbower, R. S.   675,677
Newhouse, V. L.   611,612,622
Nicol, J.   249,261
Nisenoff, M.   594,603,605
Norman, J. C.   675,677
Norris, W. T.   449,458
Notaro, J.   439,442,447,457
Notarys, H. A.   183
Novikov, Yu. N.   282,288
Nozieres, P.   78,135,136,146,147

O

Oberhauser, C. J.   486,495
Ochiai, S. I.   329,332
Ochsenfeld, R.   7,12,13,18,27,588,589,591,593
Olien, N. A.   721,733
Olsen, J. L.   423,428
Onnes, H. K.   7,8,10,11,24,27

AUTHOR INDEX

Opfer, J.	603
Oswald, B.	474,481
Ouboter, R. deB.	595,605,654,676
Owen, C. S.	564,690

P

Page, C. H.	576,584
Palmy, C.	423,428
Papp, E.	246,261
Parker, W. H.	152,565,569,584
Parks, R. D.	65,67,69,71,93,98,99,146,426
Pastuhov, A.	435,457
Patterson, A.	313,331
Paul, H.	435,457
Paul, S.	See Saint Paul
Pearsall, G. W.	348,370
Pessall, N.	358,359,371
Peterson, R. L.	556,564
Petley, B. W.	569,579,584
Pfeiffer, E. R.	374,375,386,426,428
Phillips	218
Pierles, R.	32,69
Pillenger, W. L.	217,260
Pines, D.	77,78,135,146
Pipes, P. B.	604
Pippard, A. B.	39,42,57,69,563,722,732
Pollard, E.	292
Price, P. J.	246,261
Purcell, J. R.	465,481,516,520,537

R

Rabinowitz, M.	328,332
Ralls, K. M.	291,298,304,330
Ramsay, W.	6
Pau, F.	471,472,481
Raub, E.	343,370
Rauch, G. C.	311,613
Payl, M.	347,370
Rayleigh	574,584
Reed, T. B.	347,360,370
Repici, D.	57,69
Reppy, V.	667,675,676
Reuter, G. E. H.	41,42,69,239,260
Reynolds, C. A.	7,77,146
Richards, P. L.	631,633,637,640,650,732
Richardson, R. C.	667,675,676
Rickayzen, G.	135,136,147,583,585

Ricketts, R. L.	305
Rieger, T. J.	165,172,177,183,180,748,753
Ries, R. P.	626,629
Rigamonti, A.	281,288
Roberts, B. W.	244,254,334,342,369,426
Roddy, T. J.	347,370
Rodgers, E. C.	447,449,450,451,452,453,457,458
Rogers, J. D.	516
Rorschach, H. E.	157,161
Rose, R. M.	247,249,254,261,289,291,298, 300,302,303,307,308,309,310, 316,318,319,324,328,329,330, 331,332,333,339,682,704,705, 709,739,744
Rosenblum, B.	325,326,331
Ross, J. S. H.	486,495
Rothwarf, A.	77,81,126,146,147,685,747,753
Rowell, J. M.	23,64,70,690
Ruccia, F.	435,457
Ruhl, W.	414,416,418,420,424,427,428
Russell, C. M.	426
Russer, P.	35,648,652

S

Sadagopan, V.	291
Saint-James, D.	59,70,142,148,299,301,330
Saint Paul	341
Salve, F. J. D. Jr.	427
Sands, M.	70
Sard, E.	631,647,650
Sarma, G.	142
Sasaki, W.	244,246,255,256,258,261
Sass, A. R.	614,615,622
Satterthwaite, C. B.	426,625,629
Saur, E. J.	291,360,361,370,371,426
Scalapino, D. J.	67,107,143,147,163,165,172, 176,177,178,179,180,183,284, 285,286,564,568,584,626,631, 633,640,650,682,683,684,685, 686,687,688,690,691,694,698, 732,739,748
Schaeffer, G. M.	384,387
Schawlow, A. L.	56,69
Schmid, A.	748,753
Schmidt, H.	695
Schmidt, P.	282,288,427
Schmitt, R. W.	720,721,733
Schneider, R. S.	81,147

AUTHOR INDEX

Schooley, J. F.	209,213,224,260,269,270,374, 375,376,386,405,426,428,625, 627,630,682,689,692,693,738, 745
Schrader, H. J.	569,573,584
Schrieffer, J. R.	7,18,20,25,68,71,74,78,88,90, 91,93,94,95,96,99,107,116,121, 122,125,126,128,130,135,141, 143,149,152,161,292,330,398, 403,405,426,629,683,692,737, 743,744,745,752
Schroen, W.	649,652,702
Schueler, C.	604
Schuster, H.	746,752
Schwartz, B. B.	143,548,563
Schwenterly, S. W.	667,675,676
Schwidtal, K.	572,584,700,702
Scott, N. R.	607,622
Seidel, G.	217,260
Seiden, P. E.	375,387,417,427,741,751
Sellmaier, A.	190
Senderoff, S.	439,440,457
Seraphim, D. P.	613,614,622
Seraphim, G. R. S.	631,650
Serin, B.	77,142,241,242,260,414,427
Shaktarin	486,495
Sham	683
Shamrai, V. J.	360,362,371
Shanks, H. R.	417,427
Shapira, Y.	325,331
Shapiro, S.	24,66,70,209,249,251,261,631, 635,636,637,638,639,640,641, 647,650,651,652,682,687,688, 694,697,705,707,711,738
Shchegolev, I. F.	282,288
Sheahen, T. P.	211,218,219,244,245,259
Shen, L. Y. L.	392,395,403
Shepard, L. A.	291,307
Shepelev, A. G.	246,248,261
Sherril, M. D.	418,420,427
Shiffman, C. A.	242
Shirane, G.	746,752
Shoenberg, D.	32,38,69,217,260,427
Shulishova, O. I.	292
Siebenmann, P.	377,387
Silsbee, F. B.	11
Silver, A. H.	152,156,161,563,595,605,632, 650,654,656,657,658,660,668, 671,676

Simon, Y.	752
Sinha, S. K.	399,403
Slaughter, R. J.	447,458
Smith, J. L.	190,483,486,487,490,495,496, 682,705,706,707
Smith, P. F.	536
Smith, P. H.	249,261,317,320,321,323,331
Smith, T. F.	279,288,338,362,366,368,369, 412,428
Smith, T. S.	338,339,369
Sommerhalder, R.	608,622
Sondheimer, E. H.	41,42,69,239,260
Souders, T. M.	577,578,584
Soulen, R. J. Jr.	209,627,630
Soymar, K.	246,261
Spitzli, P.	371,747,753
Spurway, A. H.	317,321,323,331
Standenmann, J. L.	353,354,371
Stans, M.	714
Stekley, Z. J. J.	462,463,481,486,495,497,500, 501,511,512,514,516,535,536, 682,702,708,709,713
Stephen, M. J.	306,308,325,330,632,651,687, 748,753
Sterling, S. A.	631,651
Stewart, W. C.	563,614,615,622
Stewart, W. D.	643,647,651
Stolan, B.	743,744
Stolfa, D. L.	660,668,676
Stoltz, O.	327,331
Stoner, E. C.	32,69
Straus, L. S.	143,148,212,244,246,249,251, 253,260,261,738,750
Strauss, B. P.	318,319,331
Stritzker, B.	414,427
Strnad, A. R.	306,308,331
Strobridge, T. R.	195
Strong, P. F.	249,261
Strongin, M.	222,241,260,288,421,428
Stuart, R. W.	185,203
Suhl, H.	287,288
Sullivan, D. B.	556,564,581,585,643,645,647
Sun, R. K.	328,332
Superata, M. A.	212,242,260
Suris, R. A.	366,372
Swenson, C. A.	428
Swift, D. A.	449,458
Swiggard, E. M.	375,387
Swihart, J. C.	569,584,626,629,691
Szklarz, E. G.	273,288

AUTHOR INDEX

T

Taber, M.	603
Taber, R.	604
Tachikawa, K.	292,460,481
Takken, E. H.	721,733
Taylor, B. N.	67,152,565,584
Taylor, M. T.	449,458
Terwordt, L.	144,148,685
Testardi, L. R.	338,367,369
Theurer, H. C.	418,427
Thiene, P.	581,585,649,652,660,662,668,676
Thomas, E. J.	142,168,169
Thompson, R. S.	428
Thullen, P.	486,487,490,495,496
Thurber, W. R.	377,387
Timmerhaus, K. D.	487,495
Ting, C. S.	287,288
Tinkham, M.	675,676
Tinlin, F.	486,495
Tiza, L.	57,69
Toepke, I. L.	426
Tomasch, W. J.	212
Toots, J.	579,581
Toth, L. W.	476,481
Trauble, H.	310,331
Tsebro, V. I.	272,751
Tsui, D. C.	738
Tsuneto, T.	164,167,183,748,753
Tung, Y. W.	377,387
Turneaure, J. P.	327,328,331

U

Uchida, M.	358,359,371
Ulrich, B.	631,651
Urban, E. W.	744,752

V

Valatin, J. G.	115,147
van Gelder, A. P.	299,330
Van Kempen, H.	628,630
Van Maaren, M. H.	384,387,427,752
van Reuth, E. C.	346,348,362,364,365,367,371,414,427
Vant-Hull, C. C.	591,605
van Vucht, J. H. N.	343,370
Vernon, F. L.	632,651

Victor, J. M.   327,328,331
Vieland, L. J.   348,370,372
Vielhaber, E.   358,359,371
Viet, N. T.   327,328,331
Vincent, D. A.   563
Vinen, W. F.   687
Voigt, H.   437,457
Vol'pin, M. E.   282,288
Von Gutfeld, R. J.   242,260
Von Minnigerode, G.   418,427
von Molnar, S.   375,387,417,427
Von Riedel, E.   632,647,651
Vystavkin, A. N.   648,652

W

Wada, Y.   626,629
Waldram, J. E.   254,255,256,257,261,563,689
Wallecka, J. D.   71,93,99,135,136,139,146
Walters, C. R.   317,320,321,323,331,536
Warburton, R. S.   673,674,675,676,677
Waring, R. K.   291
Warnick, A.   595,605,667,676
Waterstratt, R. M.   346,348,351,352,362,363,370,
                          371,384,414,427
Watson, J. H. P.   428
Webb, W. W.   38,632,651,653,667,671,674,
                          675,676,677,682,687,694,695,
                          696,698,699,703,704,711,714,
                          715,716,752
Weber, J.   604,605
Weger, M.   364,365,372,747,752
Weisbarth, G. S.   247,261
Weissman, I.   327,328,331
Wells, J. S.   571,572,584
Wenner, F.   575,584
Wernick, J. H.   291,313,325,331,338,343,369,379
Werthamer, N. R.   57,144,299,330,418,427,518,
                          632,638,639,641,647,651
Westendorp, W. F.   478,481
Weston, R.   568
Wexler, A.   625,629
Wheatley, J. C.   675
White, R. W.   338,369
Wilhelm, J. O.   334,338,369
Wilkins, J. W.   107
Willens, R. H.   316,331,346,370,371,747,752
Williamson, S. U.   301
Willis, W. D.   563

Wilman, H.	247,261
Wilson, M. N.	317,320,321,323,331,462,481,536
Wipf, S. L.	322
Witt, T. J.	579,585
Wittgenstein, F.	506,509,536
Wittig, J.	266,288
Wizgall, H.	291
Woo, J.	748,753
Wood, E. A.	343,370
Wood, J. H.	256,257,258,261
Woods, A. D. B.	395,403
Woodson, H. H.	486,490,495,496
Wright, W. H.	77
Wuhl, H.	248,261,414,427
Wulff, J.	291,298,307,311,313,318,319, 330,331,348,370
Wyatt, A. F. G.	247,261
Wyder, P.	628,630

Y

Yamafuji, K.	308,331
Yang, C. N.	149,152,160
Yangubskii, E. B.	288,292
Yaqub, M.	217,231,236,237,248,260,261, 375,386
Yoshihiro, K.	244,246,255,256,258,261

Z

Zachariasen, W. H.	273,291,336,369
Zanona, A.	433,434,456
Zavaritskii, N. V.	244,247,255,256,257,261
Zeller, G.	428
Zimmer, H.	648,652
Zimmerman, J. E.	156,161,556,563,564,581,585, 595,603,605,632,647,648,651, 654,656,657,658,662,668,671, 673,674,675,676,677,699,749, 750,753
Zucker, M.	241,242,260,414
Zvarykina, A. V.	282,288

# SUBJECT INDEX

A

A-15 compounds,
- crystallographic order in — 353-368
- crystal structure of — 275, 291, 339, 345-347
- density of states in — 365
- electron-phonon interaction in — 747
- general discussion of — 273-279
- history of — 333-368
- metallurgy of — 292, 310-313
- search for high T_c in — 353-368

Ac losses in superconductors; see Electromagnetic absorption

Ac susceptibility technique, — 212-241
- and isotropic resistivity — 226-231
- and anisotropic resistivity — 231-236

Alternators, superconducting; see Motors, superconducting

Aluminum,
- critical field curve of — 588
- enhancement effects in — 415, 416, 420
- T_c of — 294
- as a thermometric standard — 626

Anomalous skin effect — 140

B

Band structure,
- and enhancement of T_c — 391-393

BCC solid solutions,
- T_c values of, tabulated — 290

BCS theory, — 71-146
- effective electron-electron interaction in — 126-130
- gap equation in — 126
- general discussion of — 18
- and Meissner effect — 135-141
- and penetration depth — 141
- pairing interaction in — 87-93
- strong coupling corrections to — 394
- and thermodynamics — 130-135
- transition temperature in — 141

Beta-Tungsten compounds; see A-15 compounds

Bolometers, superconducting,
- noise equivalent power of — 724

Boundary scattering in superconductors — 241-247

Pages 1-428 will be found in Volume 1, pages 429-778 in Volume 2.

C

C-14,C-15 structure	278,290,313
C-16 structure	278
Cadmium,	
T_c of	294
Chemical potential,	
in BCS theory	96
in G-L theory	169-171
in nonequilibrium superconductors	171-182
Coherence length,	
in BCS theory	141
general definition of	18
in G-L theory	58,166
Pippard	40-42
Computer devices; see also Cryotrons	607-622
Constitutive equations,	
and nonlocal effects	38-42
for superconductors	26-27,36-38
Cooper's problem	87-90
Copper,	
density of	500
resistivity of	500
specific heat of	500
and stabilization of superconductors	317-324,500-504
thermal conductivity of	500
Coulomb pseudopotential,	
definition of	394
Critical current,	
definition of	11
and flux flow	306-310
and stabilization problems	461-464
in Type II materials	296,306-310
Critical magnetic field; see also Upper critical field	
anomalous	302-305
in BCS theory	134-135
defined	10,13
H_c, defined	59
H_{c1}, defined	59,297
H_{c2}, defined	59,299
H_{c3}, defined	299-300
in phenomenological theories of Type II compounds, tabulated	48-50 499
in Type II materials	297-306

SUBJECT INDEX

Critical temperature; see Transition temperature
Cryotrons, 611-617
    effective noise temperature of 724
    and Josephson junctions 617-622
    in logic circuits 614
    as memory elements 614-617
    switching time in 613,621
    thin film 611-617
    wire wound 607

D

Demagnetizing coefficient,
    definition of 31
Dielectric function,
    for free electron gas 391
Dielectric resonances in
    superconductors 390,393

E

Effective electron-electron
    interaction,
    in BCS theory 88,126-130
    calculation of, in RPA 399-402
    in jellium model 86
    in semiconductors 376-387
Effective mass of electrons,
    due to phonon dressing 740
Electrical diffusivity,
    defined 500
Electrical resistance,
    in superconductors, upper
    limit to 8-9
Electric dipole moment of ^{3}He,
    detection of, using
    superconducting techniques 603-604
Electromagnetic absorption,
    and nonlocal effects 38-42
    in magnetometer circuits 661
    in superconducting motors 484
    in a superconducting ring 153,157
    in superconducting transmission
    lines 438-444
    in Type II materials 324-329
Electron-electron interaction,
    general discussion of 77,389-393,741-744
Electron-ion interaction 401

Electron-phonon interaction,
    in A-15 compounds   747
    in BCS theory   18,78-87
    dependence of, on phonon frequency   389-396
    in strong coupling theory   394-396,741
    upper limit of   402
    and virtual phonons   79

Energy gap,
    anisotropy of   247-255
    in BCS theory   126
    definition of   72-74
    and density of states   17
    in semiconductors   379-386
    as a voltage standard   582-583

Energy gap equation,
    in BCS theory   583

F

Fermi-Thomas approximation,
    and jellium model   85

Fluctuations,
    of order parameter   683-684
    and persistent currents   9

Flux detectors; see Magnetometers
Flux exclusion; see Meissner effect

Flux flow,
    and Hall effect   308
    and pinning effects   306-310
    in type II superconductors   306-310,744

Fluxoid,
    defined   543
    quantization of   542,546,590

Flux quantization,
    discovery of   15
    and the London equations   36-38
    in a superconducting ring   151,655-659

Flux quantum,
    definition of   38
    value of   589

Flux transformer,
    use of   672-673

Flux vortex; see Vortices

G

Gapless superconductivity   17,142

Gallium,
- boundary scattering effects on T_c — 242-247
- critical field curve of — 218
- energy gap anisotropy of — 217, 244, 252-258
- foils, effect of annealing on T_c — 221
- multiple energy gaps in — 252-254
- T_c of — 212, 217, 294
- transition width in — 211, 217

Ga-In alloys,
- magnetic susceptibility of — 224

Generators, superconducting; see Motors, superconducting

Germanium,
- conduction band of — 375

Giaever tunneling; see Tunneling of electrons

Ginzburg-Landau κ parameters,
- κ defined — 58-59, 297
- κ_0, κ_1, defined — 297
- and normal state resistivity — 297

Ginzburg-Landau theory, — 57-60, 165-170
- time dependent — 167-170

Gor'kov equations — 120

Gor'kov factorization — 112-117

Gravitational radiation,
- detection of using superconducting techniques — 603-604

Green's functions, — 93-122
- and boundary conditions — 99-101
- and Nambu matrix formalism — 120-122
- and quasiparticles — 109-111
- and second quantization — 93-97
- time ordering of — 97

H

Hafnium,
- superconductivity in — 270
- T_c of — 294

Hard superconductors; see Type II superconductors

Hartree-Fock approximation — 112

Helium,
- density of — 500
- dielectric strength of — 445-447
- film boiling in — 462
- nucleate boiling in — 461

specific heat of 500
thermal conductivity of 500
HfNb,
 T_c of 290
HfTa,
 T_c of 290
Hydrogen,
 possible superconductivity in 271

I

Isotope effect, 16,76
 in Uranium 287
In-Bi alloys,
 flux pinning in 309
Indium,
 and boundary scattering effects on T_c 242-247
 critical field curve of 14,588
 and electron concentration effects on T_c 416
 T_c of 294
Iridium,
 T_c of 294

J

Jellium model 82-87
Johnson noise 628
Josephson effect, 22-24, 62-67, 149-160
 ac 65
 dc 64
 and the determination of e/h 579-580
 and flux quantization 151
 Gibbs energy in 659
 in a magnetic field 66-67
 in nonequilibrium superconductors 171-182
 in a superconducting ring 149-160, 547-550, 657
 as a voltage standard 565-582, 725, 728
Josephson equations 64, 542, 634
Josephson junctions,
 as computer devices 617-622, 726, 728
 and coupling to electromagnetic radiation 570-571, 635
 creation of suitable oxide layers in 572
 use in dc voltage measurements 569-573
 equivalent circuit for 553-556, 643

SUBJECT INDEX

fabrication of	329,662-666
noise equivalent power of	725,728
point contact	551,569,632
as a radiation detector	631-650
as a radio frequency mixer	637-642
Reidel peak phenomenon in	632
solder blob	551,569,632
stability of	572,702,705
as a thermometer	726
thin film	551,569,632
wave function in	555

Josephson tunneling; see Tunneling of electrons, Josephson effect

Josephson tunneling current,
electromagnetic field dependence of	557,566,635-642
as a function of junction bias	552
magnetic field dependence of	559,618
Reidel peak phenomenon in	647-648
spatial variation of	558-563
temperature dependence of	628

K

Kapitza boundary resistance	581

L

Landau damping	391

Lanthanum,
T_c of	294

Laves phases; see C-14, C-15 structures

Lead,
critical field curve of	14,588
phonon modes in	572
T_c of	294
London moment	544
London equations	34-38

London superconductor,
defined	42

M

Magnetic diffusivity,
defined	317

Magnetic field,
shielding against, using superconductors	590-605

Magnetic pressure,
in a solenoid	532-533

Magnetic susceptibility; see also Ac
    susceptibility technique
        general discussion of   226-241
        skin depth dependence of   222,229,237-241
Magnetization,
    and demagnetizing coefficient   31
    general discussion of   28-33
    in mixed state   21
    of Type I superconductors   19,29,33
    of Type II superconductors   20
Magnetometers, superconducting,   594-596,653-676
    as a communication device   727-728
    design of   560-563,667-676
    equivalent circuit for   661
    practical applications of   675-676
    principles of operation of   659-667
    sensitivity of   602,672,727
    and thermometry   675
Magnets, superconducting,   459-480,497-535
    applications of   520-523,726-727
    in fusion research   464-480
    magnetic pressure in   516,532-533
    mechanical fabrication of   514-516,531-535
    performance of   509-514,516-520
    quenching of   509-514
    stabilization of   461-464,500-506,
                                           524-531
    winding materials for   316-324,460-461,
                                          498-499,506-509
Matthias' rules   266,284,406-407
Maxwell's equations,
    applied to superconductors   26-28
Meissner effect,
    departures from ideal   589,591-594
    definition of   12
    derived in BCS theory   135-141
Mendelssohn sponge   32
Mercury,
    critical field curve of   14,588
    T_c of   294
Metallurgy of superconductors,
    general discussion of   289-332
    and manufacturing methods   310-316
Molybdenum,
    alloys, T_c of   409
    T_c of   294
Motors, superconducting,   483-494
    ac losses in   484
    advantages of   494,729
    performance of   490-491

SUBJECT INDEX 809

N

Nambu Matrix formalism	120-122
Nb_3Al,	
critical current in	315
critical field curve of	22, 314
fabrication techniques for	316
H_c of	316, 460
used as a magnet material	508
T_c of	409, 499
$Nb_3(Al_{.8}Ge_{.2})$,	
critical field curve of	22
H_c of	460
T_c of	460, 499
(Nb_3Al) Nb Ge,	
T_c of	409
Nb_3Au,	
T_c of	409, 348
$Nb_3(Au_{.95}Rh_{.05})$,	
T_c of	409
NbC,	
T_c of	290
Nb_3Ge,	
T_c of	409
NbN,	
T_c of	290, 409
NbN-NbC alloys,	
T_c of	290, 409
Nb_3Sn,	
critical current in	499
critical field curve of	22
density of	500
fabrication techniques for	310-313
use in power transmission lines	435-436
stabilization of	323-324
T_c of	409, 460, 499, 500
thermal conductivity of	500
specific heat of	500
upper limit to T_c in	286
Nb-Sn alloys,	
T_c of	347
NbTi,	
density of	500
specific heat of	500
T_c of	460, 500
thermal conductivity of	500
Nb-Ti alloys,	
critical current in	305, 499
fabrication techniques for	310-311

H_c of	304
use as a magnet material	506-509,510
use in power transmission lines	435-436
T_c of	290,293,499
stabilization of	317-323

NbZr,
density of	500
specific heat of	500
T_c of	290,409,500
thermal conductivity of	500

NbZr alloys,
fabrication techniques for	310
use in power transmission lines	435-436
T_c of	293

Niobium,
critical current in	303
critical field curve of	588
electromagnetic absorption in	327-329,438-442
phonon modes in	572
use in power transmission lines	436,438-442,452
pressure dependence of T_c in	423-425
T_c of	294,394

O

Off diagonal long range order	152
Order parameter,	
in BCS theory	132
and gapless superconductors	142
general discussion of	57-60
spatial variation of	174-182
Organic superconductors	269,460

Osmium,
T_c of	294

P

Protoactinium,	
T_c of	338
Pairing interaction	87-93
Paramagnetism in superconductors	298-299
Penetration depth,	
in BCS theory	141
in London theory	14,36,56,571
temperature dependence of	56,72,75
Phenomenological theories of superconductivity	25-70
Phosphorous,	
superconductivity in, under pressure	266

SUBJECT INDEX

Pippard superconductor,
    defined   42
Plasma frequency,
    in the free electron gas   391
Propagators; see Green's functions
Proximity effect,
    and enhancement of T_c 418-421
    in Josephson junctions   177-182
    and magnetometers   669-670

Q

Quasiparticles,
    recombination time of, in superconductors   583
Quenching of a superconductor   509-514

R

Radiation detectors; see Johsphson junctions
Refrigeration techniques,   185-205
    and Joule-Thomson cycle   186-188
    and Stirling cycle   192-200
    and Vuilleumier cycle   192
    and work extraction cycles   188-191
Riedel peak phenomenon,
    in Josephson junctions   632,647,648
Rhenium,
    T_c of   294
Rh-Ir alloys,
    T_c of   268
Rhodium,
    T_c of (extrapolated)   267-269
Rh-Os alloys,
    T_c of   268
Ruthenium,
    isotope effect in   287
    T_c of   294

S

Second quantization   93-96
Semiconductors,
    energy gap of   379-386
    many valley   265
    phonon modes in   376-379
    and polar phonon interactions   377-378
    superconductivity in   373-387,412,416-417

Sesquicarbide structure,
    superconductivity in   273
Shielding of electric and magnetic
    fields,
    using superconductors   587-605
Skin depth,
    anomalous   40, 237
    normal   39, 213, 224, 230
SLUG,
    Clarke type   551
    defined   671
    use in magnetometers   561
Sommerfeld condition,
    and superconducting ring   654
Sodium chloride structure,
    superconductivity in   273
Soft superconductors; see Type I
    superconductors
Splat cooling   274
SQUID,
    definition of   671
    sensitivity of   725
Stabilization of superconductors   317-324, 500-506
    adiabatic   462-463, 480
    cryogenic   461, 465
    dynamic   462-464
    effect of twisting on   323, 504-506
Superconducting devices,   688-712, 724-732
    physics of   539-563
    summarized   540-541
Superconductivity,
    discovery of   7-8
    early theories of, reviewed   13-18
    future of   682-716
    unanswered questions in   740-749
Supercooling effect in
    superconductors   52-53
Superheating effect in
    superconductors   52-53
Surface sheath,   50-54
    and H_{c_3} 59, 299

T

Tantalum,
    Coulomb pseudopotential in   392
    critical field curve of   14, 588
    T_c of   294

SUBJECT INDEX

Ta-Hf alloys,
    T_c of   293
Technetium,
    T_c of   294
Thermal diffusivity,
    defined   317, 500
Thermodynamics of superconductors,
    in BCS theory   130-135
    general discussion of   42-54
    and nonreversible effects   50-54
Thermometry,
    by critical magnetic field   217, 626
    by heat capacity techniques   625-626
    using Josephson junctions   627
    noise   628-629
    standards using T_c 217, 627
Thomas-Fermi screening length   168
Thorium,
    T_c of   294
Time dependent Ginzburg-Landau
    theory   165-170
Tin,
    critical field curve of   14, 588
    and effect of electron
    density on T_c 416
    and mean free path effects
    on T_c 415
    T_c of   294
Titanium,
    T_c of   294
Ti_3Ir,
    T_c of   348
Ti-Mo alloys,
    T_c of   290
Thallium,
    critical field curve of   14, 588
    and effect of electron
    density on T_c 416
    and pressure dependence of T_c 423
    T_c of   294
Tomasch effect   212
Transition metals,
    and alloying effects on T_c 293, 405-412
    and bond structure effects
    on T_c 271-273
    and e/a ratio effects on T_c 295
    and impurity effects on T_c 405-412
    T_c of   271, 406

Transition temperature,
- and alloying effects — 269, 405-413
- annealing effects on — 362
- in BCS theory — 127, 740
- and bond structure — 271-273
- and Curie-Weiss behavior — 282
- and Debye temperature — 337-338
- defined — 9
- dependence on average phonon frequency — 742
- and e/a ratio — 263-288, 335-342, 355
- electron density effects on — 414-417
- and electronic specific heat — 338-340
- and enhancement effects — 263-288, 389-402, 405-425
- and ferroelectric behavior — 281-282
- and ferromagnetic behavior — 282-284
- highest value of — 396, 460
- and impurity effects — 405-413
- isotope effect on — 16
- mean free path effects on — 241-247, 414-415
- and melting point — 265-277
- pressure effects on — 266-270, 280, 397, 422-425
- and proximity effect — 418-421
- of semiconductors — 384
- size effect in — 421-423
- and strain effects — 270-271
- in strong coupling theory — 394-396, 740
- tabulated values for elements — 11, 294

Transmission lines, superconducting — 433-456

Tungsten-bronze structure, superconductivity in — 280

Tungsten,
- density of states in — 393
- T_c of — 272, 294

Tunnel junctions; see also Josephson junctions
- selection of materials for — 329-330

Tunneling of electrons,
- into bulk materials — 247, 259
- introduction to — 21-24
- in multiple gap superconductors — 251-254, 738
- selection rules for — 255-259
- used as a voltage standard — 582-583

Two-fluid model,
- Gorter-Casimir — 54-57
- introduction to — 13-14

SUBJECT INDEX

Type I superconductors,
    intermediate state in   142
    magnetic properties of   19,29,33
Type II superconductors,
    critical current in   296,306-310
    discovery of   21
    electromagnetic absorption in   324-329
    flux flow in   306-310,744
    magnetic properties of   20,297-305,744
    mixed state in   20-21,59,142-143
    and Pauli paramagnetism   298

U

Upper critical field,
    defined   59,297
    enhanced by spin paramagnetism   299
    limited by Pauli paramagnetism   298
Uranium,
    isotope effect in   287
    T_c of   294

V

Vanadium,
    critical field curve of   14,588
    T_c of   294
V_3Al,
    T_c of   357-358
V_3Au,
    and long range order in   348-352
    T_c of   348
V_3Ga,
    critical field curve of   22
    H_c of   460
    pressure dependence of T_c in   424
    T_c of   290,409,460,499
$V_{4.5}Ga$,
    critical field curve of   22
V-Ga alloys,
    H_c of   344
    T_c of   344
V_3Ge,
    pressure dependence of T_c in   424
V_2Hf,
    T_c of   290
Virtual phonons in electron-electron
    interaction   79

Voltage divider,
    cryogenic   573-582
Voltage standard,
    and Giaever tunneling   582-583
    and Josephson effect   565-582
Vortex state,
    and Type II superconductivity   20-21
Vortices,
    and electromagnetic absorption   324-327
    Lorentz force on   306
    and pinning effects   306-308, 324-329, 591, 744
    pinning force on   325
    repulsion between   317
V_3Si,
    critical field curve of   22
    magnetic susceptibility of   278
    pressure dependence of T_c in   279, 424
    T_c of   290, 340, 409
    upper limit to T_c in   286
V-Ti alloys,
    T_c of   290, 293
V_3-X alloys,
    search for high transition temperature in   356-360
VZr,
    T_c of   290
V_2Zr,
    T_c of   290
$V_2(Zr_{.5}Hf_{.5})$
    T_c of   290

W

Weak superconducting link; see Josephson junction
W-Re alloys,
    T_c of   290

Y

Y-Rh alloys,
    T_c of   407

Z

Zinc,
    critical field curve of   588
    T_c of   294
Zirconium,
    T_c of   294